AVIREX U.S.A.™

PILOT - Carl Scholl
THUNDER
Pacific Princess
PBJ-1J

GILLES LHOTE & JEFF CLYMAN

COWBOYS OF THE SKY

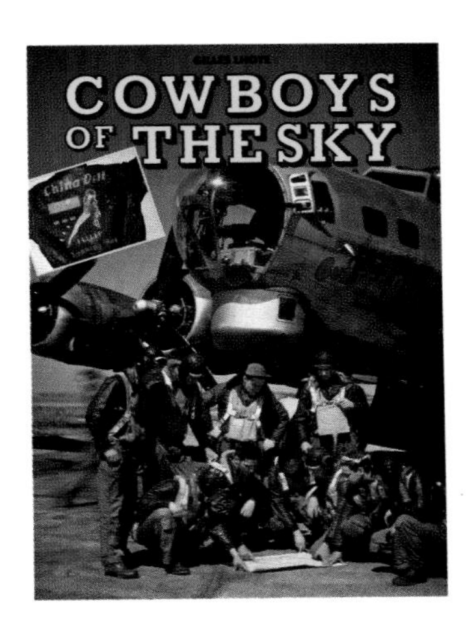

Motorbooks International
Publishers & Wholesalers Inc
Osceola, Wisconsin 54020, USA ®

This edition first published in 1989 by Motorbooks International Publishers & Wholesalers Inc., P.O. Box 2, 729 Prospect Avenue, Osceola, WI 54020 U.S.A.

Printed and bound in Italy.

Library of Congress Cataloging-in-Publication Date ISBN 0-87938-354-2

Motorbooks International
Publishers & Wholesalers Inc.
P.O. Box 2 729 Prospect Avenue
Osceola, WI 54020 U.S.A.

CONTENTS

PREFACE

A *B-3 bomber jacket of sheepskin. This model jacket was to become a legend and be immortalized by General Patton, who had used it as his favorite uniform.*

IN SEARCH OF THE AMER

"I do not consider myself a manufacturer of leather jackets, but rather a witness to history. Today, my jackets are identical to those worn by the American pilots in World War II. The original leather jackets were as strong as General George Patton's Sherman tanks or B-17 Flying Fortresses, manufactured with the heart and the guts of a victorious America. Authentic flight jackets for real heroes, both made in the U.S.A."
It was with these words that Jeff Clyman, President of Avirex, introduced himself the day I met him in his museum/office in New York. The meeting had been arranged more than a year before through André Leguen (Director of Avirex France) to help me with my first photo essay on the history of leather jackets: "The Leather of Heroes". Clyman had graciously granted me an hour of his time: but the interview ended up taking four days! An exciting trip to the depths of the hard-core universe of aviators with, as an added bonus, a fabulous trip above the skyscrapers of Manhattan in an antique North American Texan AT-6D from World War II. The book "The Leather of Heroes" appeared in

NEW YORK Journal American

AN AMERICAN PAPER FOR THE AMERICAN PEOPLE

DAILY, 5 Cents. SATURDAY, 10 Cents. SUNDAY, 15 Cents.

No. 22,260—DAILY SATURDAY, JANUARY 22, 1949 R

List 3 for Top Air Honors

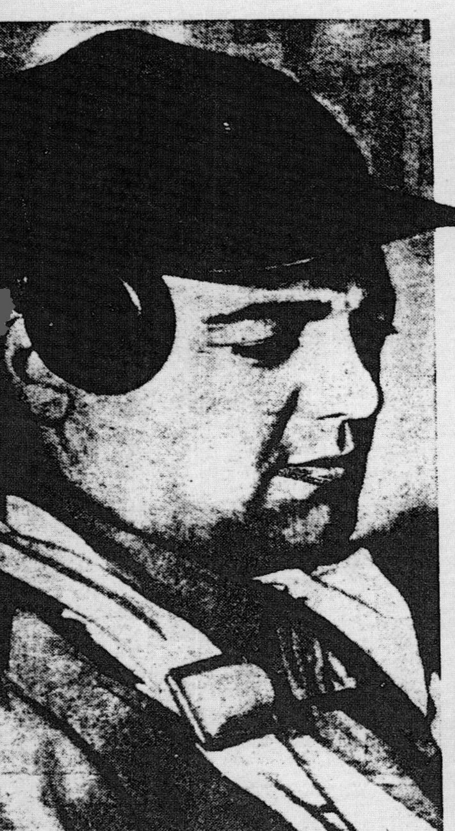

...ING LIFE IN HANDS, as airmen must, Her...isher reversed all four propellers of a plane in

EARLY FAME came to Gen. Jimmy Doolittle, when he was the first to fly blind, guided only by instruments.

IMMORTAL DAREDEVILS of the air include Captain Charles E. Yeager, first to fly an airplane faster than sound. He beat 763 mph easily.

On Saturday, June 22, 1949, the New York daily, "Journal American", published on its front page the picture of three American aviation heroes: Herbert Fisher, Jimmy Doolittle and Chuck Yeager. Three cowboys of the sky dressed in A-2 leather jackets who had "the right stuff".

ICAN LEATHER DREAM

November 1987. In February, 1988, the first batch was sold out. Patrick Mahé and Nicolas Hugnet, the directors of the Filipacchi publishing house in Paris, were already considering reprinting. But I was already imagining a sequel. In fact, numerous readers and military leather buffs wanted to learn more on the real leather of heroes. I immediately thought of Avirex. Jeff Clyman is a speed freak on land, air or sea, working at full throttle 18 hours a day at the speed of sound, jumping from his old T-6 to the most sophisticated civilian jets to go into combat in Europe, Japan, or Korea on the leather jacket front. A new meeting was arranged. Thank you, André! I explained my idea to Jeff: to go in search of the American dream; to find the survivors of the Burmese war, the war in the Pacific, the wars in Europe, Korea and Vietnam; to track the flight jacket cult to the ends of the museums in the U.S.A.; to uncover the rarest examples or the most significant; to rub elbows at air shows with the fanatics of "war birds" (the old restored World War II fighter aircraft) with whom the flight jacket is almost a religion... More or less, to rebuild in a few months of research the universe which Jeff had succeeded in

United States Air Force Museum

Dayton, Ohio

From the inside of a B-17 Flying Fortress cockpit — symbol of American power — at The United States Air Force Museum in Dayton, Ohio, to important magazines of the times such as "The Saturday Evening Post", "Life" or "Match" (the French equivalent of "Life" Magazine), we are immersed in the universe created by Avirex: that of authenticity. Poster design and photography and interior photo of USAF Museum B-17 by Dan Patterson, Cincinnati, Ohio.

creating during many long years of work. I wanted to write an authentic book — 100 % pure aviator leather.

While I was attempting to convince Jeff, he was thumbing through my book, "The Boots of America", which I had done six years before with ACLA Publishers. The whole history and legend of cowboy boots...

The ex-Arizona desert rider shot a look at the tip of my ostrich boots and said: "Okay, you can start next week, but I warn you, I don't have a lot of time to give you!" The next weekend we were on Jeff's Lear Jet 35 and arrived in Florida at Titusville Airport for the largest concentration of war birds in the world at the Valiant Air Command show. This year, the survivors of Chennault's Flying Tigers were the guests of honor of this most unusual air show. Tex Hill, one of the top flying aces of the Burmese war, was participating, and singularly impressive by his humility and simplicity.

Tracking the flight jackets continued by visits and research through the most prestigious sources; The United States Air Force Museum in Dayton, Ohio, The National Air and Space Museum in Washington, The Confederate Air Force in Harlingen, Texas, The Chino Airport Air Museum in the California desert, the airplane "bone yards" in Tucson, Arizona, and the Avirex jacket factories for the Air Forces in

© Dan Patterson

Salem, Ohio and West Virginia. Jeff had opened all the doors for me. I came back to New York with more than 50 kilos of documents and photographs — all original. Everywhere I went the idea of "Cowboys of the Sky" had intrigued: the Americans allowed me to enter into their fantastic time machine of flight jackets.

On the road again. Ten days later, Jeff accompanied me to the Sun and Fun air show in Lakeland, Florida, where "The Six of Diamonds", an acrobatic T-6 team sponsored by Avirex, were present. New connections and an impressive harvest of other documents. Jeff Clyman only granted me his complete trust after four months of work and research. During that time I had excellent contact with his wife and partner, Jacky, and partner Frank Marchese (an ex-Vietnam veteran) and the whole Avirex and Cockpit staff. A team of "super pros" living solely by and for flight jackets. Often, my friend Mark O. Greenberg, Artistic Director responsible for production of the Cockpit catalog, would lend me his photo studio at Avirex headquarters.

I felt that Jeff really opened up the day he permitted me to look through his personal archives, a loft-type room, filled with drawers from floor to celling. The boss's inner sanctum! It's in this secret hideout that the pilot turned ex-business lawyer who was bored keeps his files on his passions, idols and heroes. At

At 42, Jeff Clyman, President of Avirex, wearing his leather jacket and cap in front of the nose of a Curtiss P-40's shark teeth, has sacrificed everything for his passion for aviation and bomber jackets. A pilot of war birds and jets, he has manufactured jackets for "The Right Stuff" and "Top Gun", and works on contracts for the American Armed Forces. For more than six months, we scoured the U.S. together from East to West and North to South in search of the American leather dream.

the top of the list, General Jimmy Doolittle, a pioneer who was the first pilot to fly blind, guided only by the instruments on his panel. This same Doolittle would become famous during World War II, becoming one of the greatest AAF commanders leading the first American bombing raid on Tokyo in April, 1942. The second is another ace, a pilot, scientist and survivor of the first air bridge in history: the Hump. His name? Herbert "Fireball" Fisher, Chief test pilot at Curtiss, friend of General Chennault and notorious "cheater of death". Fisher was the first to have flown a propeller-driven aircraft and to have reversed the propeller to brake landing speed. It was he who helped invent giant, curved propellers in order to beat the world speed records. The third hero is also a living legend since we are talking about Charles "Chuck" Yeager, a top scoring World War II ace and the test pilot who on October 27, 1947, broke through the sound barrier on a prototype X-1 rocket plane. Three pilots who had "the right stuff"! Clyman, as a matter of fact, sees Fisher and Yeager from time to time. After six months of the chase, we had almost wrapped up the saga of the military jacket. The only thing missing was the ultimate piece, the rarest of the rare, the one that almost no museum nor collector owns, the leather A-1 flight jacket invented in 1925 and standardized on November 7, 1927.

We all know that this legendary animal, the ancestor of the A-2 flight jacket, existed because we had seen

it on Cary Grant in the movie "Only Angels Have Wings", but where could we get the original pictures? It was Lindbergh, another Clyman hero, who put us on the right road. But of course! Why not go to The Charles Lindbergh Museum on Long Island at Roosevelt Field, where the first cowboy of the sky had taken off on May 21, 1927, for Le Bourget. Jeff and Jacky's partner, Frank Marchese, went with me to Roosevelt Field, where the curators of the museum welcomed us so warmly that we spent two days there. Forty-eight crazy hours digging through kilometers of treasures! Finally, we uncovered a number of documents on the olive green A-1 in lambskin, closed by buttons since the zipper was only invented in 1928. We had found the missing link! The only thing left to do was to transcribe my notes and cassettes, write the book, and sort through 74 kilos of documents and hundreds of "Life", "The Saturday Evening Post" and "Match" magazines from 1939 to 1945. The best years for the leather of heroes and the cowboys of the sky. Welcome to the flight jacket fanatics' club!

*

Special thanks: I wish to thank particularly here the following friends who made this crazy idea become reality: Jacky and Jeff Clyman, Frank Marchese, and André Leguen, Mark O. Greenberg, Patrick Mahé ("Paris-Match") and Pierre Ménager (Surplus Neuilly). ■

New York,
Bryant Park
Pay to the Order of Charles
Twenty five Thousand NO/100
Payable in funds
at New York Clearing

PIONEERS OF THE SKY

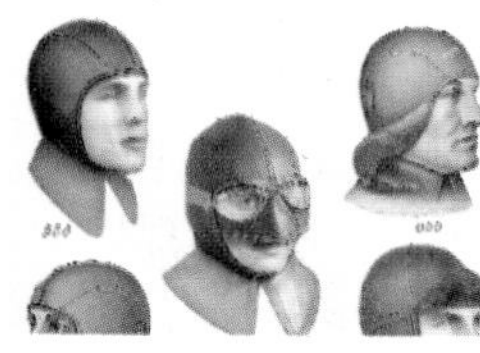

No conquest, even that of the moon, created such delirium as the conquest of the Atlantic by airplane 61 years ago. During the night of May 21, 1927, the single engine Ryan 220 H.P. belonging to Charles Lindbergh, "The Spirit of St. Louis", landed at Le Bourget, near Paris, France, amidst a delirious crowd. Against impossible odds, the conquering knight had left 33 hours earlier from Roosevelt Field on Long Island, near New York. He had conquered the Atlantic. A legendary achievement! Lindbergh had defied death jokingly in saying about his contemporaries before crossing the Atlantic: "They take me for a somewhat crazy cowboy..."; a cowboy of the sky, who was also a pioneer of the American Air Postal Service and who wore better than anyone else the historic leather aviator jacket "summer A-1". We found incredible pictures of the first A-1 jackets at The Lindbergh Museum at Roosevelt Field. The A-1 was the ancestor of the A-2 flying jacket of World War II pilots.

Jeff Clyman and Frank Marchese tracked down the check which Lindbergh received after his exploit: one for $25,000 signed on June 17, 1927, and drawn on the Bryant Park Bank of New York. In remembrance of Lindbergh, Avirex has created linings and labels based on this check. One of the pioneers of aviation, Ruth Law, training on one of the first Wright airplanes. A couple of pioneers in 1920. Note the suede shirt with patch pockets. One of the ancestors of the A-1 jacket?

It is December 17, 1903, at 10.35 in the morning; two Americans, Orville and Wilbur Wright, inventors of bicycles, are busy with a funny-looking machine on the beach of Kittyhawk, North Carolina. Orville Wright lies down on this machine made of cloth, wood and wire. A motor backfires. The machine, which has no wheels, glides along its gliding rail, takes off and stays in the air by itself for a distance of 36 meters at a speed of about 15 kilometers per hour. Aviation has just been born, and with it the age of aviation clothing... From 1903 to 1911 the first pioneers of the sky braved rain, snow, the cold and oil with very rudimentary materials wearing generally a leather coat, motorcycle goggles, football-style helmets and long leather gloves which came up to their elbows. We would have to wait till the last days of 1911 for a real interest to develop in creating pilots' clothing. As of 1913, catalogs of flight accessories for men and women would start appearing in the U.S. and in Europe including coats and leather helmets, as well as the first aviation sunglasses. The First World War was to be the catalyst that forced Army labs to address the problem of aviation clothing. When the U.S. entered the war in April, 1917, the manufacturing for the small "Army Air Arm", part of the Army Signal Corps, could not face the demand and asked for help from the French and British allies as well as commercial apparel companies. More often than not the first American pilots who participated in the war equipped themselves on their own. Yet a basic uniform did exist: A 3/4 coat in brown leather, double-breasted and with a belt. This first "flight coat" had an officer's collar, a breast pocket and two other flap pockets. In addition to the leather helmet and glasses, the American pilots wore khaki or olive-green uniform pants cut like riding pants which were worn inside high leather boots. Their hands were protected by heavy leather gloves.

After the war ended in 1918, the aviator "style" evolved as progress was made in the world of aeronautics. The better the airplanes performed, the higher and faster they flew, the more the clothing engineers sought to quickly develop innovative uniforms. The most important steps were taken in 1920, the year in which "air mail" started. The pilots who took mail from one end of the United States to the other, in all weather conditions and all types of situations, in open cockpit airplanes, had worn tough and practical uniforms. We should note that at this time the Germans had a slight technical advantage over the Allies and Americans. The German ace, the "Red Baron" Manfred Von Richtofen, was already wearing a short 3/4 black leather coat, fur lined and very well cut, as early as 1915. But America would soon overcome that advance by attacking, as of 1921, on two fronts: that of heated flight suits and lined flight suits. The first heated flight suits resembled long underwear with a top part sewn with electric wires. The pilot would plug his flight suit into a socket in the back of the plane. Later, these heated flight suits would become more sophisticated and would also be used in World War II. In April 1921, Lieutenant J.A. MacReady beat a height record pushing his airplane (with open cockpit) to above 40,000 feet. To fight the cold he was wearing an experimental flight suit which was to evolve into the Type B-1 model nicknamed "The monkey suit". In fact, the monkey suit was not at all made of monkey skin but of the fur of Chinese dogs. The skin was turned inside out and so the pilot wore the fur inside. The "monkey suit" was warm and fairly comfortable to wear but it soon became the aviator's nightmare: the dogs were flea infested and

The Illinois legislature has voted against state enforcement of prohibition, but it is not believed residents of Chicago will notice an appreciable difference

The Weather

THE DAILY OKLAHOMAN

152263

LINDBERGH'S OWN STORY

France Wildly Elated As Lindbergh's Flight Ends

EFFORT TO CUT LEVEE RESULTS IN GUARD CALL

GREAT PERIL IS SEEN

Index

PACIFIC TRIP MAY BE NEXT FOR LINDBERGH

ELATED AT SUCCESS

The Modern Magellan

Gas and Oil Cost $175 For Lindbergh's Flight

OCEAN HOPPER QUIT FARM FOR CAREER IN AIR

LIKES TO BE ALONE

Slim Missouri Flier Finishes Successful Hop From New York

Lindbergh Index

SECOND PARIS HOP DELAYED

FRENCH FLIERS TO BE HONORED

Money Overlooked By Flood Looter

The front cover of a daily American newspaper, "The Daily Oklahoman", which published a picture of Lindbergh with the caption: "Gas and Oil Cost $175 For Lindbergh's Flight".

the suit had a terrible odor so it was impossible to wear. Yet the "monkey suit" would continue to be made until 1930 and we found pictures of Lieutenant Benjamin S. Kelsey wearing a "monkey suit" as late as 1931 at Mather Field. As of 1923, the uniform research specialists of the Army Air Arm started to work on a project for garments that would be short, light and functional, made of leather. Two years of testing were needed before the first leather jacket,

the leather A-1 jacket, was perfected in 1925 and standardized in 1927. This date is important in the history of the leather flight jacket since the A-1 is really the ancestor of the A-2.
This flight jacket, which is almost impossible to find today, was really the favorite uniform of all American pioneers. Its characteristics were not those of a traditional flight garment: in fine olive green lamb, its inspiration was rather "sport fashion". The collar,

Charles Lindbergh and his wife, Anne Morrow, posing in front of "The Spirit of St. Louis". In this picture, which is courtesy of The Lindbergh Museum, Lindbergh is wearing his summer A-1 and his laced aviator boots.

cuffs and waistbands were made of wool knit. The A-1 closed with buttons since, as has been said, the first zippers were only invented in 1928. Two patch pockets, low in the front, were also closed with buttons. The A-1 was so popular that some pilots still wore them at the beginning of World War II. Nevertheless, this flight jacket was declared out of style and useless on September 29, 1944.
Today it is even harder to find an original A-1 leather jacket than it is to find a living brontosaurus: no museum or private collector has any. As we have seen, this rare flight jacket appeared in a short episode in the movies, Cary Grant having worn it in "Only Angels Have Wings". The A-1 would be a favorite of Charles Lindbergh and Jimmy Doolittle. Considered an elegant piece of clothing, the pioneers of the sky would not decorate it with any painting or inscription as they did the A-2 flight jacket a few years later. On the contrary, the A-1 was carefully respected and cherished. It was usually worn only with official unit insignia and with riding pants pushed down into shiny leather boots. In 1927 catalogs, we have found traces of the A-1, which was also manufactured by civilian companies and which strangely resembled a "luxury item" (60 years ago it cost more than $60) worn with a white shirt and a tie during flying rallies.
The pilots would have to wait until 1928 to be correctly uniformed. At that date the Army Air Corps pilots were issued with Type B-7 flight suits, lined A-2 boots in Shearling with a small pocket nicknamed mocassins, Type B-3 leather helmets, goggles for low altitude flying or face masks for higher altitude tests. The Type S-1 parachute also became practical and lightweight. Other innovations of the era included the A-4 gloves called "aviator two-finger gloves" because two of the fingers were enclosed in the same space. In 1928, the summer Type A-3 flight suits with patch pockets on the knees and breasts were issued and continued to be issued in limited quantities until 1944. But, in fact, 1928 was an important date in the history of the equipment for pilots of the Army as well as pilots of the mail delivery service because at this time the zipper was invented, permitting better closing of flight suits such as the B-7 and B-8 models. The invention of the zipper was an incredible revolution which announced the coming of the important American aviation era and the radical evolution of flight equipment.
It is important to note that, at the same time the Army was developing flight clothing, the U.S. Navy was actively pursuing its own program of clothing development. The first Navy leather flight jackets closely resembled their Army counterparts, but the Navy also developed lightweight cotton flight jackets along the same lines as the Army A-1, since the new aircraft carriers Lexington and Saratoga were operational in the Pacific ocean by the end of 1928.
While the Army continued to use elegant lambskin for its A-1, the U.S. Navy opted for goatskin leather, which was more resistant to wear and fairly light in weight. The early Navy flight jackets had knit cuffs and collar, but still closed with buttons, and it wasn't until the early thirties that the classic U.S. Navy "G" series jackets were adopted, with zipper closures. But the U.S. Naval Air Arm was still in its infancy in 1929-30, and most research effort was spent on Army Air Corps fight clothing, since the Army was considered the predominant military air arm.
All the ingredients were ready for the birth of the most mythical aviator jackets: the A-2 flight jackets and the B-3 Bomber jackets. Nothing could stop the arrival of the "legends"! ■

***O**pposite: Two pages from an old catalog of aviation clothing and accessories dated 1928 show us the cream of the crop of the English and American aviation pioneers. Lindbergh in his "monkey suit". The crew of the "Question Mark" wearing A-1 jackets: James H. Doolittle, Art Goebbel, etc. Above: John H. Mears, nicknamed in 1926 "the modern Magellan" in his magnificent A-1 before beating Graf Zeppelin's hot air balloon endurance record for a round-the-world flight. Rooseveld Field during the year of Lindbergh's achievement, as seen by the 1st Squadron flying in formation. "The Six Aliens" (actually Army Air Corps pursuit pilots) posing in front of their Boeing P-26 Pursuit Plane wearing leather B-1 jackets — equally impossible to find today —, the ancestors of the B-3 bomber jacket. Note: Jeff Clyman actually has an example of the B-1 in his private collection.*

1. Sir Alan and Lady Cobham, on the Thames, about to start their 20,000-mile-around-Africa flight in the "Singapore."
2. The inauguration at Miami, Fla., of the first international passenger and mail service between the United States, Cuba and West Indies. Photo shows—L. to R., Assistant Secretary of Commerce for Aeronautics, MacCracken and Mrs. MacCracken, Miss Earhart, Postmaster General New and S. I. Glover, 2nd Assistant, in charge of air mails.
3. La Nina, the first international service air mail plane to arrive at Havana.
4. Lieut. Ben Eielson and Capt. Sir George Hubert Wilkins, who flew over the Arctic, at City Hall, New York.
5. Commander Byrd—A close-up.
6. L. to R. Oscar F. Grubb, mechanic, and Capt. Hawks at Los Angel cessful trans-continental flight to New York—18 hours, 22 minutes, be 36 minutes.
7. Lieut. James H. Doolittle, U. S. Army Pilot, winner Schneider cup race.
8. Art Goebel, first prize winner of $25,000 Dole Air Derby, Oakland to H
9. The Southern Cross.
10. Crew of the Southern Cross flight, San Diego to Australia. L. to R. Smith, Lieut. Ulm, James Warner.
11. Amphibian Biplane arriving in New York with mail from the Ile de took off 500 miles at sea.

18 19

G 4
192-

International Newsreel
Underwood & Underwood

1. **Colonel Lindbergh** and Mr. Harry Guggenheim dressed for winter flying.
2. Crew of the "Question Mark," after the world's record of six days in the air, leaving ship at the Metropolitan Airport.
3. Miss Elinor Smith, 17, who made an endurance flight of 13 hours, 17 minutes.
4. Miss Amelia Earhart, first woman to successfully fly the Atlantic, arriving at Southampton, England. L. to R. Lew Gordon, mechanic, Miss Earhart, Wilmer Stultz, pilot.
5. Lady Mary Heath, famous English aviatrix in her Moth plane, in which she made her round trip solo—London to South Africa—welcomed by the Mayor of Miami, Fla.
6. Miss Bobby Trout, 21, of California, who made an endurance flight of 12 hours, 11 minutes.

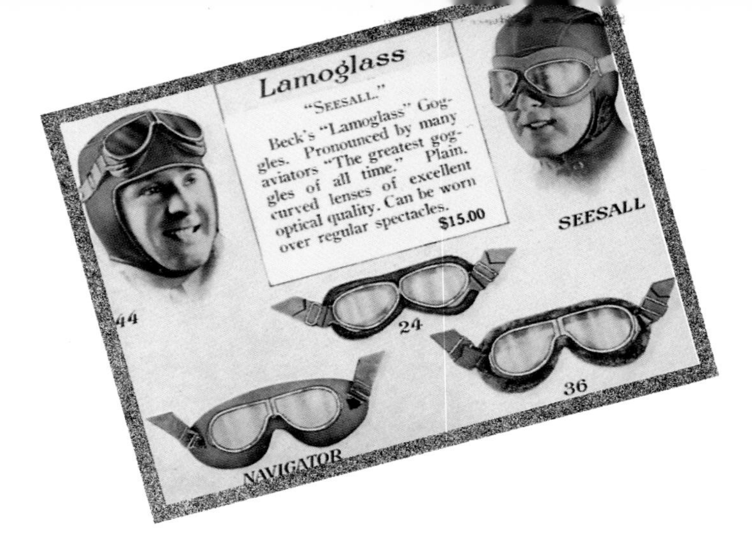
Lamoglass
"SEESALL."
Beck's "Lamoglass" Goggles. Pronounced by many aviators "The greatest goggles of all time." Plain. curved lenses of excellent optical quality. Can be worn over regular spectacles.
$15.00
SEESALL
44
24
36
NAVIGATOR

Opposite: These four aviators belong to the "Dare Devil" Squadron. They were wearing leather jackets Type A-1 that we see here in an excellent close-up. More than a simple combat or high-altitude flight jacket (like the A-2 or B-3), the A-1, created in 1925, was conceived in the spirit of a "sports" jacket. This jacket remained available in limited quantities until the end of World War II on September 29, 1944. Already a hell of a good look! At left: Glen H. Curtiss, the father of the legendary Curtiss airplanes. Below: A ravishing female pioneer of the skies in a flight suit. Above: The overview of New York in 1934: The brand-new Empire State Building face-to-face with the first wave of modern Army "Keystone" Bombers.

The 18 pilots posing for the classic group shot in their summer A-1 jackets are part of the famous 77th Fighter Squadron, whose patch represented a poker hand (four sevens plus an ace), and who also made their mark on the European front during World War II. In memory of these elite aviators, Avirex produced a 77th Fighter patch. Above: Jimmy Doolittle, an authentic American hero, proudly posing in his olive green A-1. An aviation pioneer, Doolittle would be the first pilot to fly blind in a closed cockpit solely on instruments. During World War II, he would become one of the most famous Commanders leading the first bombing raid on Tokyo in April of 1942. Opposite: The first U.S. Mail pilots and William H. Bleakley, another pioneer, in a lined leather flight suit from 1928 nicknamed "blanket lined, winter flying suit".

G-18

Opposite: For many years, Jeff Clyman and Frank Marchese would seek out information on Charles Lindbergh, one of their heros. They would find such treasures as the $ 25,000 reward check for the crossing of the Atlantic and this identity card dated 1925 and belonging to one of the most celebrated pilots of the U.S. Army Air Service. In homage to Lindbergh, this document has become an Avirex lining. Above: Lindbergh when he worked for the St. Louis Postal Delivery Service. In front of the airplane is Irvin Clover, Postal Chief Assistant, and Frank H. Robertson who flew with Lindbergh on the St.Louis/Chicago route. Lindbergh, the cowboy of the sky, in his leather A-1 in front of his favorite airplane, "The Spirit of St.Louis".

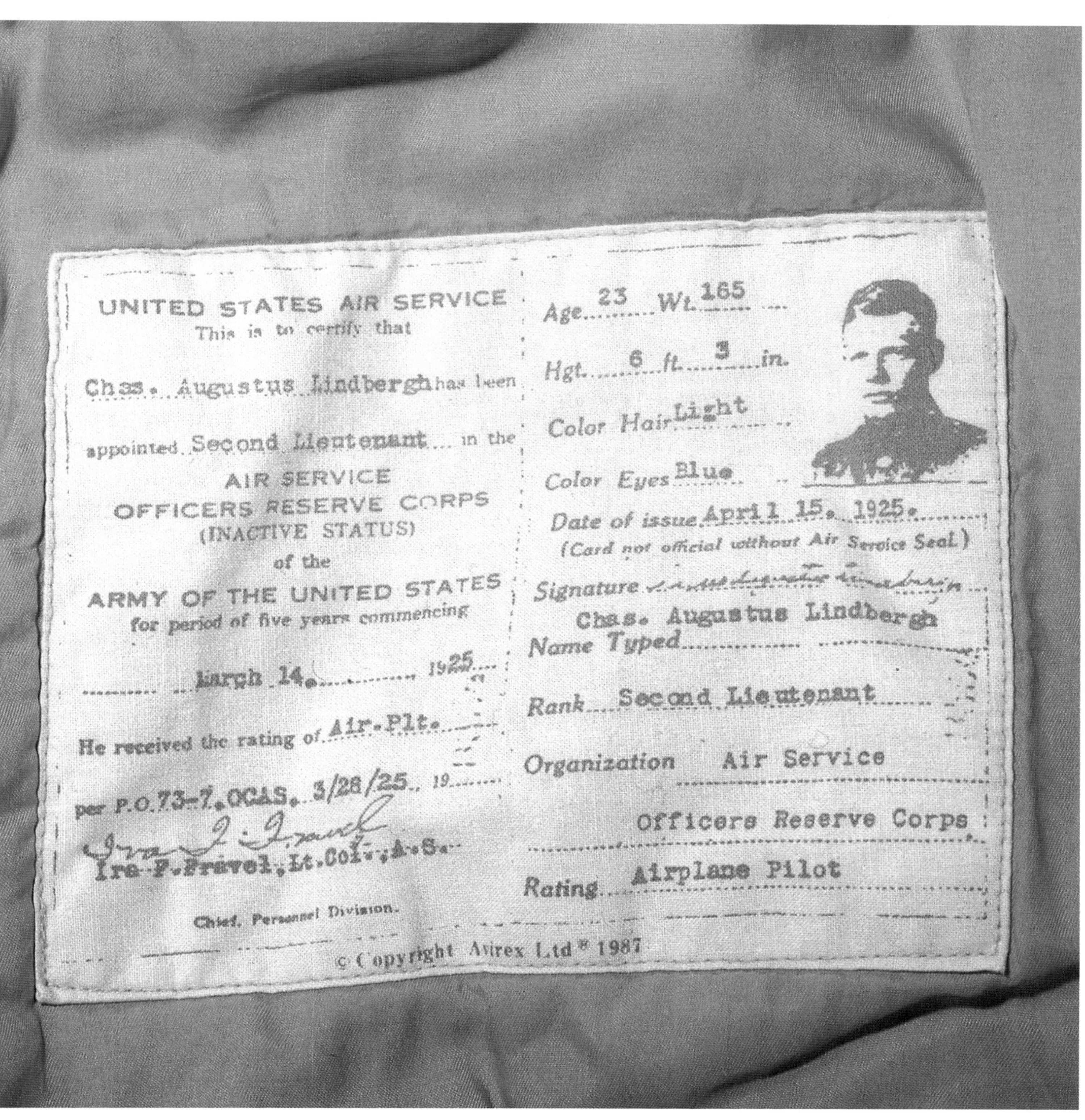

UNITED STATES AIR SERVICE
This is to certify that

Chas. Augustus Lindbergh has been

appointed Second Lieutenant in the

AIR SERVICE
OFFICERS RESERVE CORPS
(INACTIVE STATUS)
of the
ARMY OF THE UNITED STATES
for period of five years commencing

March 14, 1925

He received the rating of Air. Plt.

per P.O. 73-7, OCAS. 3/28/25. 19

Ira F. Fravel, Lt. Col., A.S.

Chief, Personnel Division.

Age 23 Wt. 165

Hgt. 6 ft. 3 in.

Color Hair Light

Color Eyes Blue

Date of issue April 15, 1925.
(Card not official without Air Service Seal.)

Signature

Name Typed Chas. Augustus Lindbergh

Rank Second Lieutenant

Organization Air Service

Officers Reserve Corps

Rating Airplane Pilot

© Copyright Avirex Ltd® 1987

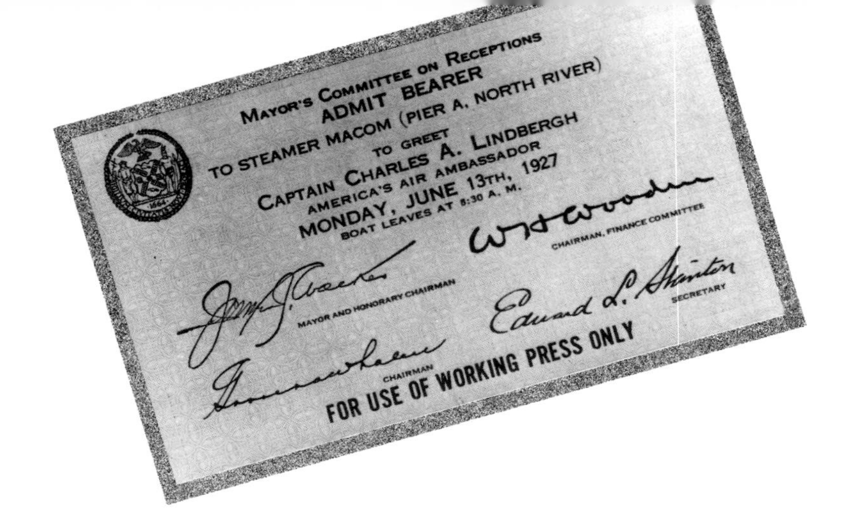
Mayor's Committee on Receptions
ADMIT BEARER
TO STEAMER MACOM (PIER A, NORTH RIVER)
TO GREET
CAPTAIN CHARLES A. LINDBERGH
AMERICA'S AIR AMBASSADOR
MONDAY, JUNE 13TH, 1927
BOAT LEAVES AT 8:30 A. M.
CHAIRMAN, FINANCE COMMITTEE
MAYOR AND HONORARY CHAIRMAN
SECRETARY
CHAIRMAN
FOR USE OF WORKING PRESS ONLY

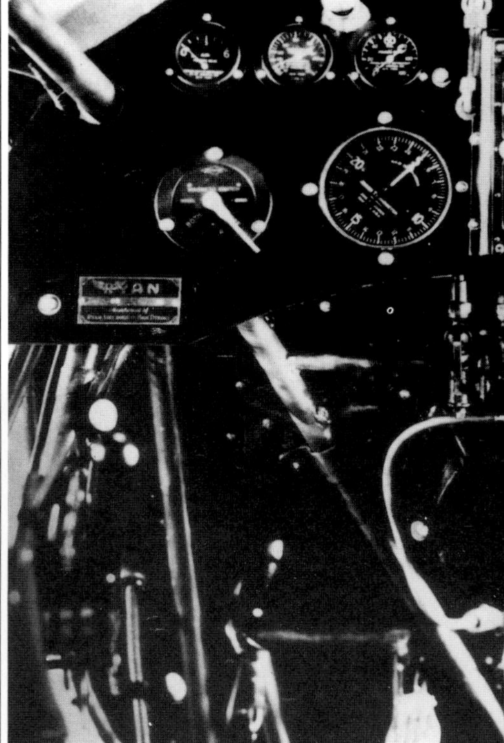

Opposite: "The Spirit of St. Louis" just before its takeoff from Roosevelt Field, a 220-horsepower airplane manufactured by Ryan Airlines of San Diego with a Wright air cooled radial engine. The airplane when stripped did not even weigh a ton. It was made of wood and canvas and Lindbergh's seat was in cane. The "Spirit" did not have a parachute, nor a gas gauge, nor a radio. Lindbergh, about six feet three inches tall and weighing 165 lbs, only took with him five sandwiches and a liter of water. Above: The instrument panel aboard "The Spirit of St.Louis". Upon arriving at Le Bourget airport, the people of Paris, during this magical night of May 21, 1927, consecrated Lindbergh a "cowboy of the sky". As soon as the news reached the States, all America exploded with joy. Such a spontaneous delirium would never be seen again until man's landing on the moon. Top left: The press pass permitting access to a reception in Lindbergh's honor.

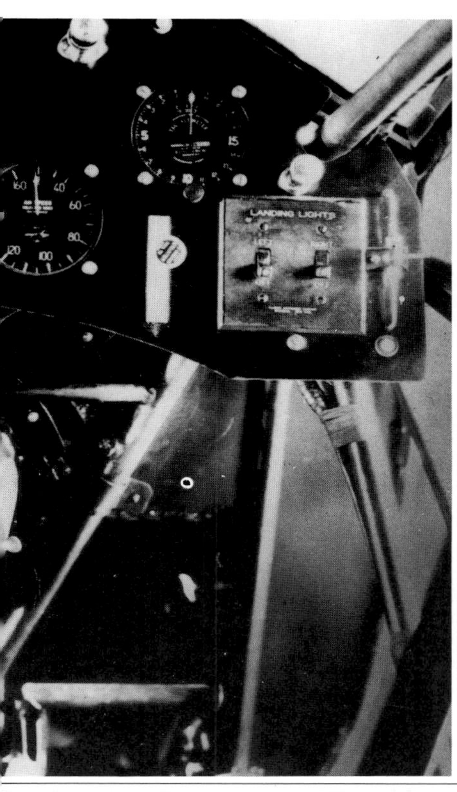
LANDING LIGHTS

N-X-211
RYAN

ERICKSON
SAN DIEGO.

Left: Charles Lindbergh at a time when he was a specialist in aerial acrobatics. He had two airplane accidents and escaped death twice. Below: Lindbergh is congratulated by the Mayor, Mr. Thatcher, at Albany Airport. Opposite: Lindbergh in 1942. America had been at war for a year and Lindbergh was a reputed Nazi sympathizer who made pro-German speeches... But even so, the controversial hero was to go into battle unofficially (as a civilian) against the Japanese in the Pacific, where he completed more than 40 missions and rehabilitated himself by shooting down two Japanese fighters. He died on August 26, 1974, on the Hawaiian island of Maui, where he had retired from the world. He was 72 and died from cancer. Opposite at left: The leather "Charles Lindbergh" style coat which Avirex created to commemorate the 60th anniversary of the Atlantic crossing.

On December 7, 1941, after the Pearl Harbor debacle, millions of young Americans were living happily in the most powerful and mythical country in the world. They all wanted to marry starlets and resemble John Wayne. Many were to leave for the war in the Pacific or in Europe. Many were to die. But not those who thought their leather bomber jackets, in horsehide or sheepskin, would make them invincible... On their backs they painted pinups and on the cockpit of their airplanes, victory slogans. These jackets were conceived for war. Today they are the stuff of dreams. Avirex has perpetuated the U.S. tradition in manufacturing these "authentic bomber jackets"... as strong as Patton's Sherman tanks, or the Boeing B-17 Flying Fortresses. Opposite: Carl Scholl of Aero Trader Inc. at Chino Airport in California has had his G-1 Type leather jacket painted with the same motif as is painted on the nose of his B-25 bomber — a delightful naked Pacific princess leaning against a bomb. Three war paints: 'Miss Gee Eyewanta", on the back of a pilot of a B-17, inspired by an illustration from "Esquire"; "Hell's Henchmen", appearing on an A-2 jacket of a member of a B-17 crew; and "Fitch's Bandwagon", also painted on the A-2 of a pilot of a B-17. The bombs represent the number of missions accomplished.

WAR PAINTS

WAR PAINTS

This superb flight jacket, a Navy G-1, on which is hand-painted a female "Lonesome Angel", naked and sporting little wings, on an American flag, is part of an important private collection owned by Jeff Clyman. It belonged to a pilot of an F-4-F Wildcat. A beautiful piece in perfect condition, size 44. "Heavenly Body", "Miss B. Haven" and "Rosie's Sweat Box" are also hand-painted on the backs of A-2 flying jackets having belonged to members of Flying Fortress crews. Note that all of these girls were very scantily clad and pleasantly voluptuous, copied from pinups in men's magazines of that era. We will also find the same sources of inspiration on the nose sections of different fighters and bombers of the period.

This flight jacket Type A-2 hand-painted with a "Wild Child" motif belongs to the Avirex collection. Its design was inspired by the flight jacket of an American aviation ace, Lieutenant Colonel Robert W. Waltz. Waltz's A-2, made of horsehide, had the words "Wild Children" and 50 bombs showing that the pilot had completed 50 missions, as well as 12 swastikas, indicating that Waltz, belonging to the 390th Bomb Group, had downed 12 German airplanes. The two other "Grin 'n Bare It" flight jackets have the same additional designs and the same number of bombs; only the pinup is different. The two pilots belonged to the same B-17 crew but had different tastes in women.

WAR PAINTS

This hand-painted flight jacket is also part of Avirex's latest collection. It is inspired by two things. On one hand it pays homage to the 380th Bomb Group. As for its reference to the Bengal Wing Officers' Club, it refers to the "Hump" pilots who flew over the Himalayan Mountains to supply the American forces surrounded by the Japanese in China. The Hump was the first air bridge in history. The goods or materials used to arrive by cargo ship from the U.S.A., and were offloaded in India at Calcutta. There, the 10th Army Air Force based in Assam flew the cargo at altitudes above the highest peaks in the world, over 30,000 feet. These three flight jackets were worn by pilots of the 12th and 15th Air Forces fighting on the North African and European fronts.

W A R   P A I N T S

HEADQUARTERS
380th BOMB
GROUP
AVIREX
BENGAL WING
OFFICERS CLUB
1945

RMY AIR

COWBOYS OF THE SKY

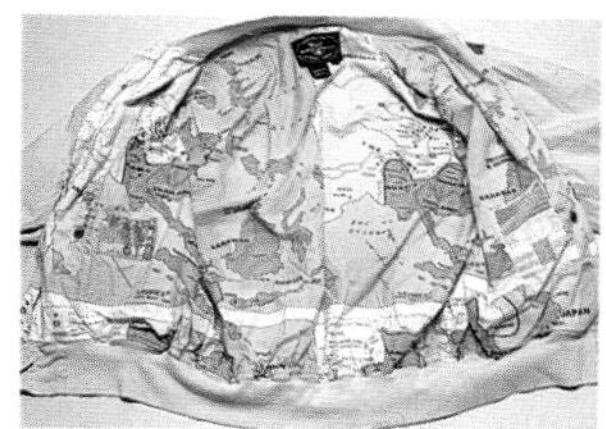

Cowboys of the Sky, the title of this chapter and of this photo essay, was inspired by the fantastic saga of the legendary "Flying Tigers" led by the mythical Claire Lee Chennault. As of 1937, Chennault, and later, in 1941, the first American pilots of the American Volunteer Group, which included aces such as Tex Hill, Ed Rector, and Chuck Older, used to do battle in China at the side of General Chiang Kai-shek against the invading Japanese armies. In Europe, on September 1, 1939, Germany started World War II by invading Poland. On September 3, 1939, France and England declared war on Hitler... On December 7, 1941, the Japanese "Zeroes" bombed Hawaii and destroyed the American Fleet and a part of the American Air Force. After the disaster of Pearl Harbor, the United States finally mobilized and came into the war. The American Volunteer Group with Chennault then became the "Flying Tigers" known in Chinese as the "Fei Hu".

On all fronts, the A-2 and G-1 leather flying jackets, and Types D-1, B-3 and B-6 sheepskin flight jackets, were to become immortalized by the airmen who wore them as well as by famous generals such as MacArthur, Patton and Montgomery. Built like tanks or B-17 Flying Fortresses, these unequaled Army Air Force jackets were to revolutionize the world and permanently influence fashion.

***More** than 40 years after having been painted on flight jacket leather, the logo of the Army Air Force and its colors still have as much impact. The wings of victory. Above: The waist gunner of this Flying Fortress did not economize on bullets. He wears a Type F-2 heated flight suit, a Model ANH16 winter helmet, and did not even take off his goggles to aim at the attacking German fighters. Below: A real cowboy of the sky. Lieutenant John Carpenter of the 118th Tactical Recon Squadron based in Lulang, China, in 1944. Carpenter was a P-51 Mustang pilot. It was his squadron that was depicted in the 1988 Steven Spielberg movie classic "Empire of the Sun", attacking a Japanese airfield deep in China.*

FLYING TIGERS

"You the Flying Tigers, you sought combat in the sky for our country against enemies who were 30 times more numerous. You have written a remarkable page in the history of that war. Your image will remain engraved in our memory." It is by this quote borrowed from General Chiang Kai-shek that the Valiant Air Command at Tico Airport in Titusville, Florida, welcomed in March 1988 the survivors of the Flying Tigers. Among them was Tex Hill, "Chennault's Supreme Warrior". During the four days of this fabulous air show, which was a reunion of innumerable "war birds", the real ex-Tigers were the show stoppers. After more than 47 years, these extraordinary older men, in perfect physical shape, piloted their favorite airplane, the indomitable P-40 Warhawk. Humble, almost embarrassed at being the center of attraction, Tex Hill signed autograph after autograph and posed for all pictures. Let's get in the old time machine and go back in time to better understand that heroism of these American "adventurer" pilots who wore the most prestigious and rare of flight jackets. A hero's flight jacket ornamented by an insignia representing a flying tiger designed by... Walt Disney!

Right: A Curtiss P-40N Warhawk taking off from one of the runways at Tico Airport, Titusville, Florida, in March 1988. A jewel in perfect flying condition. Below left: Claire Chennault with one of the old crew members of his acrobatic flight team "The Three Men on a Flying Trapeze". Above: A still shot taken from the film "God is My Co-Pilot", starring Dennis Morgan, in 1944.

Miss Josephine

As of September 29, 1930, the laboratories of the Clothing Branch of the Army Air Corps started working on a project for a new "summer" leather flight jacket. Their research was based, naturally, on older leather jackets such as the A-1. On May 9, 1931, the A-2 flight jacket was finally standardized. No one at the time could imagine that this leather jacket would revolutionize the world and become the "cherished child" and "second skin" of the American Air Force pilots and of generations of people worldwide who would idolize it.

At the time, the A-2 had a fabulous look and a great cut. This "ultimate" flight jacket was made of dark brown horsehide, lined with light brown spun silk, and sported an "officer's collar" with snaps underneath, epaulets, two patch pockets and knit cuffs and waistbands. The name of the pilot, printed on a piece of lighter leather, was stitched on the breast, and the jacket was emblazoned with hand-painted unit insignia on the right and left breasts.

In April, 1934, Major E.L. Hoffman, Chief of the Equipment Branch, noticed after numerous tests that sheepskin flight suits called Alaskan suits were ideal for the air mail and Army pilots. In order to make the sheepskin more resistant and impervious, the outside of the skin was hand-coated with a special brown paint mixed with a lacquer. When this finish dried and got older, it cracked, showing the lighter "crust" of the skin underneath. These antique sheepskin flight suits were made and shipped in two parts: The B-3 winter flying jacket and the A-3 winter flying trousers. It was in this way that another flight jacket was born, the famous B-3 bomber jacket whose fame would be publicized years later over the skies of Europe by the gunners of B-17 Flying Fortresses, B-24 Liberators, and by Generals Patton and Montgomery, who always wore them. Just as with the A-2, the B-3 would appear everywhere. Thanks to Avirex Ltd., these two flight jackets are today, more than 50 years after their creation, fashion classics destined never to go out of style.

But the first American pilots who would baptize these A-2 and B-3 flight jackets in combat were the famous American Volunteer Group (AVG) Squadrons who would begin the fight in November of 1941, in Burma and China against the invading Japanese. Claire Chennault, who had already been in China since 1937 as Technical Advisor to General and Mrs. Chiang Kai-shek was the foundation on which the AVG was based.

Claire Chennault, a descendent of French Protestants from Louisiana, is one of the most mythical figures in American aviation. Born in Flatlands near Waterproof, Louisiana, Chennault grew up on his parents' farm in open country. He was a simple man of nature, a fisherman and a hunter who was to be exposed to the elements all his life. This was no doubt the reason why the Chinese would nickname him affectionately "old leather face". When he was a young man, he was a school teacher in Franklin parish before joining the Army Signal Corps in 1917. Having become a pilot, Chennault used to get the Army aviation Chiefs of Staff on his back because of his revolutionary theories. He affirmed that the age of chivalry in the skies was gone and that only

On August 10, 1942, Claire Chennault's photograph was on the cover of "Life" Magazine, which carried a long article about the "visionary soldier" and the life of the Flying Tigers in their headquarters in Chungking. Fame thus finally came to Chennault, who had been the bugbear of the traditionalist American generals.

results were important — that one had to catch the ennemy off guard, shoot him down, and take off on other missions. Very soon Chennault prevailed because of his exceptional aviation qualities, including his excellent strategies, contact with his men, and his "visionary" enterprises. His superiors hated him and so at 40 he was still only a captain. He left the Army to start with two of his friends, John Williamson and William McDonald, an acrobatic air team called "The Three Men on a Flying Trapeze". From 1932 to 1936, these three would acquire a legendary reputation. The key to their show was named "The tailspin of death". The three airplanes were connected by short cords attached to the wings, and the pilots did a series of tailspins. They were already wearing magnificent A-2 leather jackets and proved to the military the immense possibilities for air ambush. At that time China was falling apart under the blows of the Japanese. Desperate, Chiang Kai-shek sent, in 1936, a reconnaissance mission to the U.S.A. to ask for help. The Chinese envoys were astounded by "The Three Men on a Flying Trapeze" during an air show in Florida... John Williamson and William McDonald were hired on the spot; Chennault hesitated a few months before joining them in China.

Events there went from bad to worse: The Marco Polo front collapsed. Indo-China fell into Japanese hands. As of 1940, Chennault decided to recruit an army of mercenary pilots.

Chiang Kai-shek approached the United States Government for help and President Roosevelt was overwhelmingly receptive. Convinced that the United States should soon join the fight against Japanese

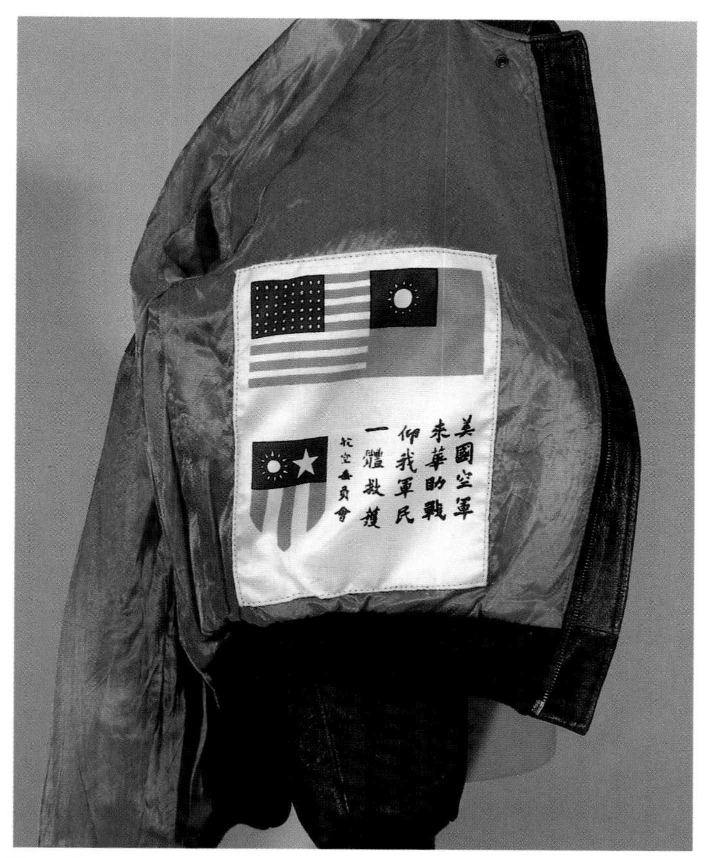

The "Flying Tigers' Jacket", created by Avirex in honor of Chennault and his "Cowboys", with the red lining showing that the pilot was an ace who had downed more than five enemy planes, and the American and Chinese flags and "blood chit" which made the pilots recognizable as allies of the Chinese.

and German attempts at world domination, but hamstrung by restrictive "neutrality" laws imposed by an isolationist Congress afraid of the responsibility of "foreign entanglements", Roosevelt and Chennault concocted a way for China to obtain modern aircraft and professional U.S. military pilots. By executive order, Roosevelt arranged that Army and Navy pilots who chose to do so could "volunteer" for service with Chennault, work as civilians for Camco, a private corporation, and upon completion of their contract be reinstated at their original rank, plus any seniority, if they chose to re-enter the U.S. Armed Forces. The planes were Curtiss P-40B Tomahawks, purchased by Britain in 1940 but declared surplus to their needs since the British found the P-40 Tomahawk inferior in both high-altitude performance and armament. By June, 1941, British Spitfires were armed with 20mm cannon and the P-40B only had two .50-caliber and four .30-caliber machine guns.
The planes were "returned" to the possession of the U.S. Government. The result of all this maneuvering was the creation of the First American Volunteer Group.
Since the U.S. was not officially at war, Chennault arranged, by a secret agreement, for the U.S.A. to deliver him 100 Curtiss P-40 airplanes and also discreetly recruited the best pilots from the Navy and Air Force. Important aces such as Tex Hill, Ed Rector, Jack Newkirk, R.T. Smith, Bob Prescott, Greg "Pappy" Boyington, and Chuck Older were also to find themselves in China, landing at Chungking headquarters with leather jackets and cowboy boots. That is why the Chinese nicknamed those who were to become their saviors "cowboys of the sky".
In July, 1941, Dutch cargo ships left the port of San Francisco with about 100 U.S. pilots and more than 150 mechanics aboard. These men were listed as missionaries. Strange missionaries who, on a stop in Natal, Brazil, had cowboy boots made to order by local shoemakers. The legend is on its way! All these mercenary pilots, not yet thirty, were to collect fabulous salaries from the Chinese Government: $800 a month plus a bonus of $500 for each Japanese plane shot down.
In the Chinese camp at Chungking, the first AVG pilots lacked everything, especially flight clothing. The Chinese would try to alleviate this situation by buying from the American Government tens of thousands of dollars' worth of G-1, A-2, B-3, and M-445 jackets as well as flight goggles, blankets, etc. Immediately, Chennault's aces did marvels in battles with odds of 30-to-1. But his tactics and strategy, added to the tough P-40 qualities, worked miracles. At this time Chennault already predicted the Pearl Harbor disaster but, as usual, no one listened to him. At the end of 1941 the AVG would become the "Flying Tigers" with the logo designed by Walt Disney. The Flying Tigers' jackets would become symbols. To prevent the American pilots from being massacred by the Chinese population during bailout or accident, Chennault and Chiang Kai-shek came up with the idea of placing a Chinese Nationalist flag on the back of the leather jackets, under which would be written the following in Chinese: "I am an American pilot. I am an enemy of the Japanese. I cannot speak your language. If my airplane is destroyed, kindly take care of me and take me back to the closest friendly military base. My government will compensate you." This message was to become the "blood chit" and became legendary. In addition to the "blood chit" the Tigers' jackets would also have the respective squadron patches, which at the time numbered three: Hell's Angels Third Squadron, Adam and Eve First Pursuit and the Panda Bear Second Squadron... The Flying Tigers' jackets were impressive enough but the ultimate distinction was to have a red lining. This color signified, "purely and simply, that the pilot was an ace who had shot down more than five enemy airplanes." We found one of these jackets in a military leather store in Santa Monica, California. The A-2 jacket, made of horsehide in a large size, was in perfect condition with blood chits and AAF insignias, as well as a Hell's Angels insignia. The owner wanted approximately $4,000 for it...
In memory of both Jeff Clyman's father, Martin Clyman, who had fought in China and Joseph D'Anna's father, who had fought in India, and in commemoration of General Claire Chennault, one of Frank Marchese's heroes, Avirex manufactured numerous collections with Flying Tigers' logos. These Avirex jackets, of which there are about ten, are all somewhat different, and are perfect reproductions of those worn by the first members of the AVG Group, then the Flying Tigers, and finally by the 23rd Fighter Group which replaced them in 1942.
But we have not finished with Chennault and his Tigers, who would continue to be known in China, in the war in the Pacific, and finally during the first air bridge in history, the "Hump". ■

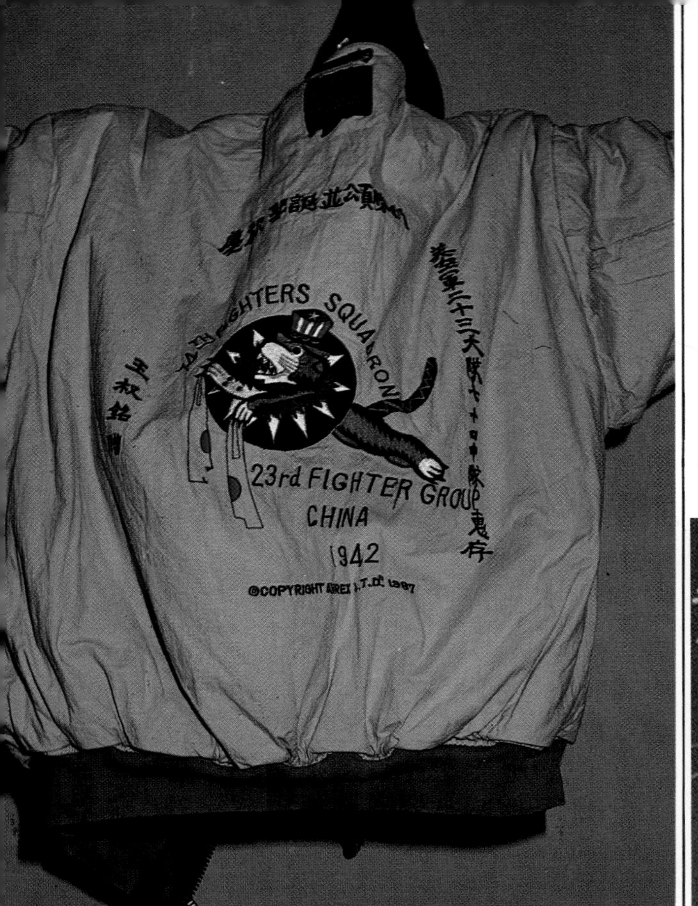

It was Chennault and his Flying Tiger

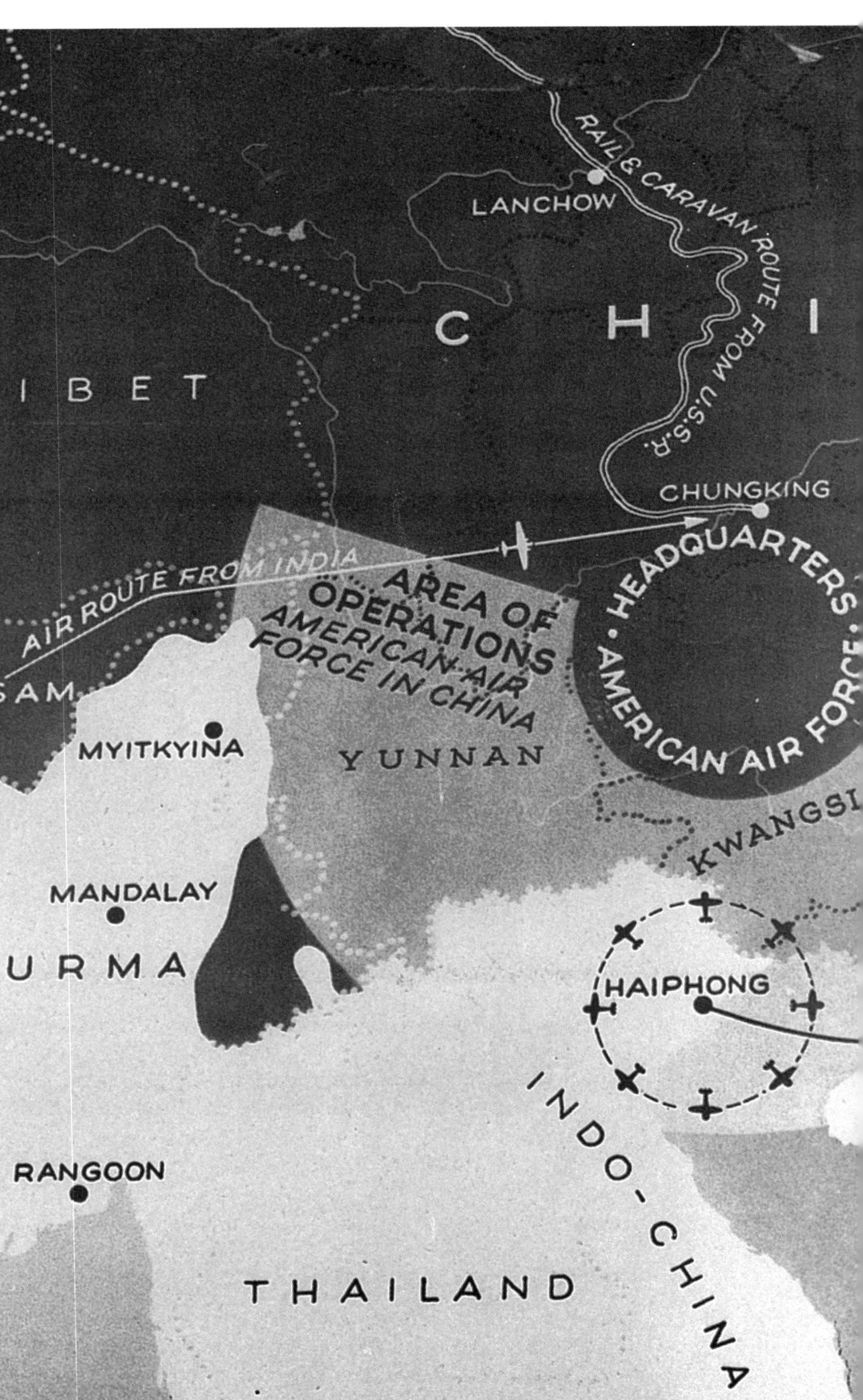

The map showing the general headquarters of the Flying Tigers at Chungking, practically encircled by the Japanese. Chennault and his men were supplied by aircraft which had to fly to China over India and the Himalayan Mountains. This dangerous supply operation was nicknamed "Flying the Hump". The lining of an Avirex jacket with an emblem of the 23rd Fighter Group. The 23rd Fighter Group was a unit of the 14th Air Force. Above: The emblem of the 75th Fighter Squadron, "The Flying Sharks", which was part of the 23rd Fighter Group fighting in China. Top right: General Chiang Kai-shek with Claire Chennault on the left. Centre right: The Curtiss P-40 Warhawk which succeeded the Tomahawk, and which was the only airplane at the time in China which could fight against Japanese Zeroes and win. Bottom right: Three American war correspondent photographers on the frontier of China and Burma. The second from the left is Syd Greenberg of the U.S. Army, who was a personal photographer of General Chiang Kai-shek, and who provided this picture.

vho gave us the title "Cowboys of the Sky".

***F**ar right: The back of a hand-painted A-2 Avirex leather jacket. This example is sumptuous. A magnificent "Chinese Doll" in a negligee is surrounded by a war environment typical of the era. The bombs represent a number of complete missions, the flags, the number of Japanese airplanes shot down, and the ships those that had been sunk. The "blood chit" reminds us of the Flying Tigers. Above: A group of "Tigers" in front of the legendary sign indicating the frontier between China and Burma. The sheepskin jackets are B-6s. At right: An advertisement for the Allison engines which powered the Curtiss P-40 airplanes, in a 1942 edition of "Life" Magazine. Below: The adventures of Chennault's Flying Tigers have inspired numerous authors and writers — one of the most authentic accounts is that written by Malcolm Rosholt, "The Days of the Ching Pao". "Flying Tigers" by Larry M. Pistole is equally important. His uncle was a member of the Flying Tigers.*

han a Pilot's Shoulders"

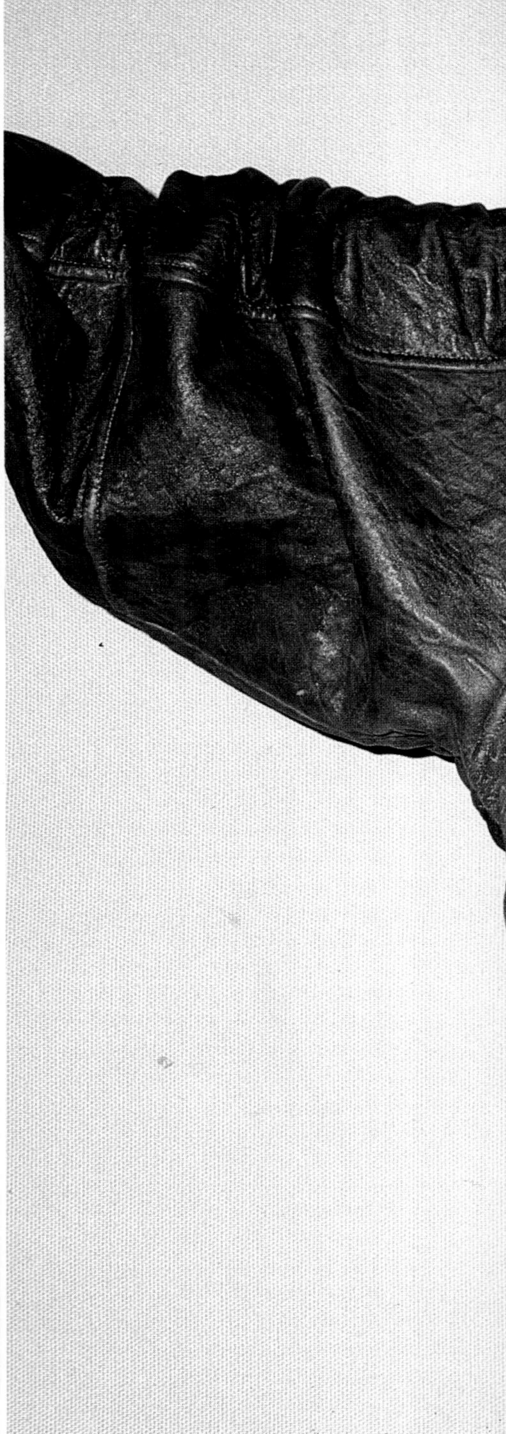

China Doll. An A-2 jacket hand-painted with a graceful woman scantily dressed in a war environment...

Milton Caniff immortalized the Tigers and the First Air Commandos in the cartoon series "Terry and the Pirates".

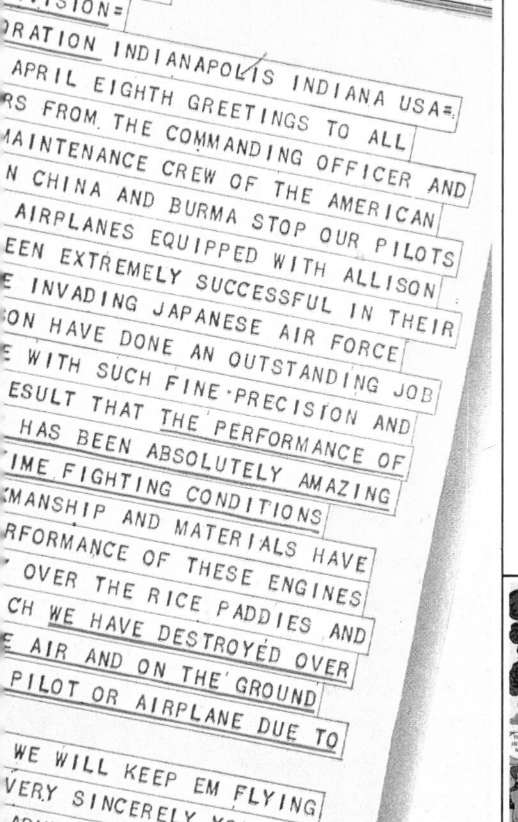

CIATE SUGGESTIONS FROM ITS PATRONS CONCERNING ITS SERVICE

TERN UNION
BLEGRAM

NEWCOMB CARLTON, CHAIRMAN OF THE BOARD — J. C. WILLEVER, FIRST VICE-PRESIDENT

SYMBOL	
LC	Deferred Cable
NLT	Cable Night Let
	Ship Radiogram

N APR 21 NLT
DIVISION=
RATION INDIANAPOLIS INDIANA USA=
APRIL EIGHTH GREETINGS TO ALL
RS FROM THE COMMANDING OFFICER AND
MAINTENANCE CREW OF THE AMERICAN
N CHINA AND BURMA STOP OUR PILOTS
AIRPLANES EQUIPPED WITH ALLISON
EEN EXTREMELY SUCCESSFUL IN THEIR
E INVADING JAPANESE AIR FORCE
ON HAVE DONE AN OUTSTANDING JOB
E WITH SUCH FINE-PRECISION AND
ESULT THAT THE PERFORMANCE OF
HAS BEEN ABSOLUTELY AMAZING
IME FIGHTING CONDITIONS
MANSHIP AND MATERIALS HAVE
RFORMANCE OF THESE ENGINES
OVER THE RICE PADDIES AND
CH WE HAVE DESTROYED OVER
E AIR AND ON THE GROUND
PILOT OR AIRPLANE DUE TO
WE WILL KEEP EM FLYING
VERY SINCERELY YOURS=
ARMY, COMMANDING AVG.

IS BY TELEGRAPH OR CABLE

NOTE: The American Volunteer Group, popularly known as the Flying Tigers, paint their planes as shown here to simulate the snout of the deadly tiger shark. The Allison engine, which permits that lethal shark-nosed streamlining, also powers thousands of other United Nations craft.

Chennault and the first members of the American Volunteer Group, later the Flying Tigers, have been the source of infinite inspiration. All the ingredients were present for the creation of a fabulous surreal image: the personality, the courage, the authenticity and the aura of Chennault, the unequalled look of the Curtiss P-40 airplanes with the painted shark mouths, and the unparalleled look of the ornamented or painted leather jackets. All during the war, the Allison Aircraft Engines division of General Motors maintained an exceptional advertising campaign of which the "Tigers" and the P-40s were the heroes. Camel Cigarettes followed in their footsteps by publishing cartoons in which the Flying Tigers and their jackets played a star role. Colonel Phil Cochran, a colleague of Chennault, was the hero of cartoons in the U.S. by the important cartoonist Milton Caniff depicting the fictional characters "Terry and the Pirates". Chennault was also the hero in France of a cartoon series called "Buck Danny", invented by the well-known reporter Jean-Michel Charlier. Above: The back of an A-2 Avirex leather jacket manufactured in honor of Chennault's American Volunteer Group.

The heroes of the war in Burma were also the Curtiss P-40s with their shark-mouth noses.

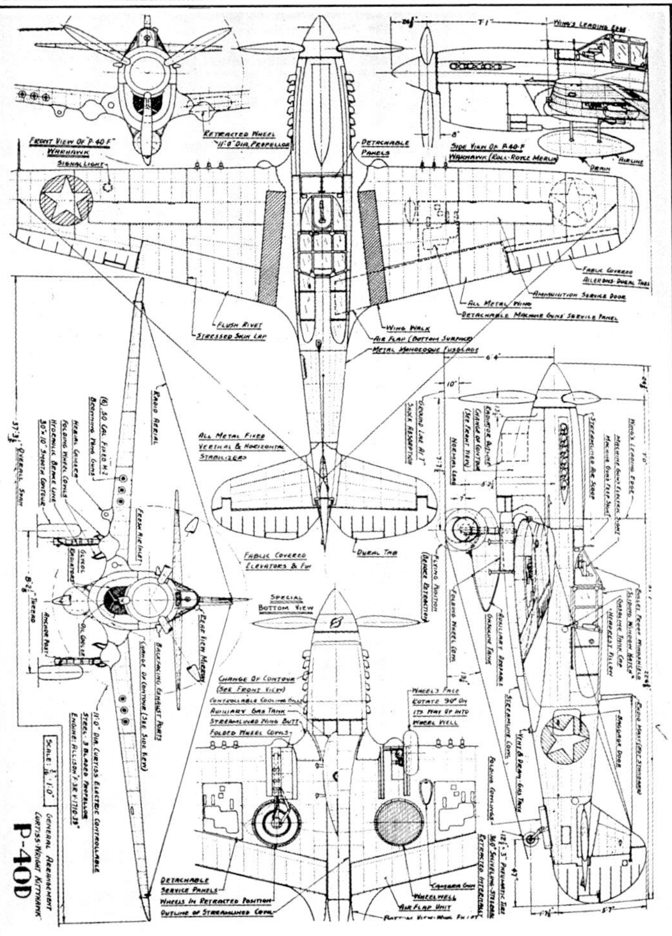

***During** the first six months in China, Chennault's Curtiss P-40s piloted by the best aces of the Flying Tigers shot down more than 300 Japanese aircraft. Holding one of the most prestigious records of World War II, the P-40 Tomahawk was neither as fast nor as maneuverable as its Japanese adversary, but it had other qualities such as exceptionally tough construction, ability to keep flying in spite of damage, considerable fire power, and incomparable "dive" speed. Chennault, who was a master tactician, had advised his Tigers to isolate their enemies and to attack them in groups of two and three. Far left: Two P-40 Warhawks in formation during an air show, the first one piloted by Bill Dodds, Jeff Clyman's friend and a Six of Diamonds pilot, photographed by noted photographer Philip Makenna, and, at left, plans of this exceptional airplane which transformed the face of war in the skies. Top right: an authentic sheepskin bomber jacket of the A.V.G. with the famous "blood chit" sewn on the back. Above: The back of an Avirex A-2 painted in memory of the movie "The Flying Tigers".*

This superb A-2 in horsehide is a museum piece. It belongs to Jeff Clyman. It was worn by a member of the first American Volunteer Group which arrived in China in 1941. The insignia sewn on the shoulder represents the famous 14th Airforce symbolized by a tiger with two minuscule wings. It is Clyman's favorite jacket, which he practically never lets out of his sight. Below right: Two original photographic documents, the complete 3rd Squadron of the A.V.G., "Hell's Angels", in front of their "good luck" P-40, and briefing scene of the Tigers at the headquarters in Kunming. Above: Three badges showing the historical evolution of the Flying Tigers. Above left is the insignia of the Hell's Angels Squadron of the original AVG. At center is the insignia of the China Air Task Force, immediate successor to the AVG when it was incorporated into the Army Air Forces on July 4, 1942. Above right is the emblem of the 23rd Fighter Group, the final incarnation of the old AVG.

the emblem of the Flying Tigers was originally designed by Walt Disney. . .

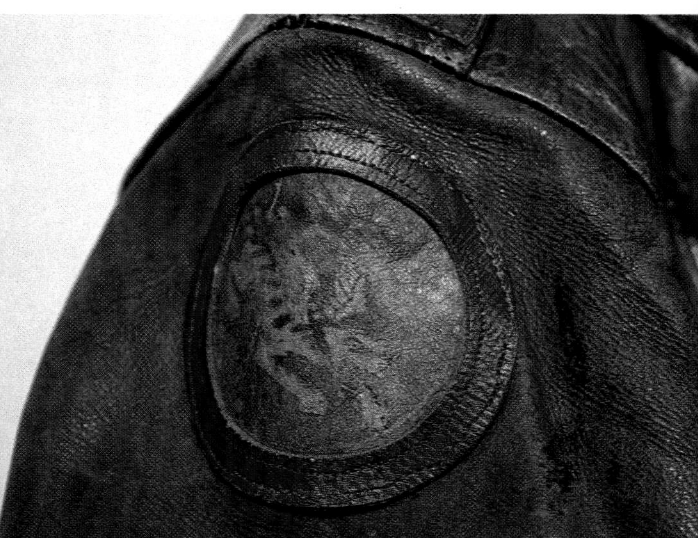

Colonel "Tex" Hill, a survivor of the original Flying Tigers, was one of Chennault's right hand men, a real war hero "made in the U.S.A."!

Far left: In March, 1988, at Tico Airport in Titusville, Florida: during the meeting of the Valiant Air Command "war birds". The guests of honor were the surviving Flying Tigers, among them the ace Tex Hill. We organized a photo with Jeff and Jacky Clyman and their two sons, Scott and Shawn, The Six of Diamonds flight team (sponsored by Avirex and the Cockpit catalog) and Tex Hill, the ultimate surviving pilot of the war in Burma and Chennault's flying right hand man. Above: Again, David L. "Tex" Hill — 47 years ago as he was preparing to get into his P-51 at Chungking. At left: Tex Hill today with his Flying Tigers' cap and his Disney-designed Tigers' badge. He is a humble giant who, in the image of his prestigious chief, is an authentic hero "made in the U.S.A.". Below: A secret document signed by Chennault directed to the 14th Airforce headquarters. In this letter, Chennault indicated the latest Japanese positions and the goals of his future missions.

SECRET

HEADQUARTERS FOURTEENTH AIR FORCE
A. P. O. #627, C/O POSTMASTER
NEW YORK CITY, NEW YORK

SECRET
By Authority of:
C. G., 14th A. F.
24/2/44 .... [initials]
Date Initials

25 February 1944.

SUBJECT: Letter of Instruction.

TO : Alert-Net Liaison Team, Changsha, China.
(Attention: Lt. Rosholt).

1. All indications point to a reshuffling of enemy units in the Yangtze Valley, west of Anking. Increases in enemy strength are reported in the Wuchang, Yochow, and Nanchang areas. New concentrations are reported at Yangsin and Tayeh. Increased air activity has been reported at enemy fields in the Wuhan area and at Nanchang. It is believed that it is within enemy capabilities to mount offensives against Changsha, both from Yochow and Nanchang.

2. You will immediately direct inquiries to the Headquarters of the Ninth War Area to ascertain disposition and strength of enemy concentrations in the following areas: Wuhan cities, Wuchang-Yochow Railroad, Yochow, Tayeh, Kiukiang, Hukow, Kiukiang-Nanchang Railroad, and Nanchang. You will report this information by radio immediately on receipt to this Headquarters and to the Headquarters of the Sixty Eighth Wing.

3. Also, you will report by radio immediately to this Headquarters and to the Headquarters of the Sixty Eighth Wing on enemy ground patrol activity and enemy aerial reconnaissance in the areas listed in Par. 2.

4. Also, you will report by radio immediately to these Headquarters all information in the location and character of enemy supply depots and dumps, transit camps, and troop movements.

C. L. Chennault
C. L. CHENNAULT,
Maj. Gen., U.S.A.,
Commanding.

cc - CO, 68th Composite Wg,
APO #430.

"I am an American pilot, enemy of the Japanese and friend of China. . . If my airplane is destroyed, help me to escape. . ."

Far left: This photograph is a very rare document. It was taken by Tiger ace R.T. Smith in China in December 1941. It represents five Curtiss P-40s in formation as they were returning to the headquarters in Chungking. It is also signed for Jeff and Jacky Clyman by many of the surviving Flying Tigers. Above: Another important document: To simulate forces which were superior to reality, the Flying Tigers had built fake P-40s in balsa wood to fool Japanese reconnaissance flights. Left: The "blood chit" sewn on the back or on the lining of the bomber jackets. Translated, it means: "I am an American pilot. I cannot speak your language. I am an enemy of the Japanese. If my airplane is destroyed have the kindness to protect me and to direct me to the closest allied post. The government of my country will recompense you." A cover of "Life" Magazine dated August 16, 1943, asking the question: "How strong is Japan?"

PACIFIC

PEARL HARBOR

On December 7, 1941, two waves of 353 fighters, torpedo planes and light bombers took off from Japanese aircraft carriers under the command of Vice Admiral Nagumo, heading for the Hawaiian island of Oahu. After having crossed the 240 miles which separated the Japanese carriers from the American forces in the Pacific, the Mitsubishi A-6M2 "Zeroes", Aichi Type 99s and Nakajima Type 97s went into attack. At 7.55 a.m., the first bombs exploded on Hickam Field, Wheeler Field, and Ford Island Naval Air Station. In a few minutes the American Air Forces, equipped with all new Curtiss P-40 and Grumman F-4-F Wildcat fighter planes and B-17 Flying Fortresses, were wiped out. At the same time another wave of Japanese airplanes attacked the famous "battleship row" in Pearl Harbor where the pride of the U.S. fleet were docked, among them the famous "USS Arizona". The disaster was total: 18 ships sunk or severely damaged, 188 airplanes destroyed, and 3,581 military and auxiliary personnel killed. That day, the Pacific fell into the hands of the Japanese and America was thrust into the heart of World War II. This surprise attack, described as "The most audacious in all of history", had been in preparation for months by Admiral Isoroku Yamamoto, Supreme Commander of the Japanese Imperial Fleet, yet it had been predicted in advance by the clairvoyant Chennault and a few spies of the U.S. Secret Services.

Far left: These pictures are taken from the American super production, "Tora! Tora! Tora!" filmed in 1970 by Twentieth Century Fox with Joseph Cotten, Jason Robards, and Soh Yamura. They show the takeoff of the Japanese Zeroes from the decks of the aircraft carrier "Akagi" and the destruction of the American P-40s at Hickam Air Force Base. At left: An ad from that time intended to help mobilize the Americans in the war industry and a patch made by Avirex in honor of the U.S. soldiers in the Pacific: "Remember Pearl Harbor".

Soon after Pearl Harbor, the U.S. mobilized its armed forces and the race for armament and equipment started. Very quickly, the usual official suppliers to the Army and Navy were out of production capacity and the U.S. Army Clothing Branch had to call on private companies to subcontract the manufacture of jackets, flight suits and uniforms. Hundreds of different companies then began working for the armed forces. Some of the most famous were: Aero Leather, Cagleco Sportswear, Roughwear Clothing, Arrow Leather, Inc., H.L.B. Corp., Perry Sportswear, Edmund Church, Osterman, Air Comfort, Air Associates and Aviators' Equipment.

The American factories were overflowing with horse, goat, and sheep skins, the raw material necessary in the manufacturing of the Type A-2 summer leather jacket, the Type B winter jacket, and the Type F flight suits, both heated and nonheated. In northern areas fighter and bomber crews flying at high altitudes wore leather garments, and those in the Pacific wore lighter cloth jackets and flight suits which were more appropriate to such warm climates. The more popular cotton flight suits worn by the Army and Navy aircrews were the ANS-31 flight suits with two patch pockets on the breast, and a further two on the lower leg. They were belted at the waist and tightened at the ankles with two adjustable straps. These flight suits in olive and light beige cotton were very light and practical. Today they are sold commercially by Avirex as well as the similar textile jackets. But these uniforms were still not ideally suited to all environments and so, as of 1944, other models of flight suits were introduced, more appropriate for warm climates: the K-1, in cotton, for the Pacific, which was very lightweight, and the similar L-1, in very supple light wool gabardine, for Europe. These two models were equipped with waist tabs and, for the first time, at least six zipper pockets. Once again the look and the functional aspects of these Army clothing items were to have an impact on the world of fashion, and on work clothing. But to return to the Flying Tigers...

America had just entered World War II, having suffered the terrible defeat at Pearl Harbor when, in China, Chennault and the success of his AVG softened the blow of the Hawaiian disaster on the public and gave renewed courage to both America and Europe, crushed as this latter was by the seemingly unstoppable German Army. Between Christmas and New Year 1941, three weeks after Pearl Harbor, Chennault's Tigers fought in Burma in the skies above Rangoon against an enemy ten times superior in numbers. The American pilots shot down 75 Japanese airplanes, losing only six Curtiss P-40s and two pilots. Chennault's strategy had worked 100%: his men attacked the Japanese Zeroes three-to-one at high speed and left immediately to go after another target.

From that point on, the whole world watched China and Burma, where a handful of heroes wearing leather jackets with insignia patches and "blood chits" faced and fought off the "invincible" Japanese armada, at last showing that the enemy could be smashed by courage and American know-how...

On September 17, 1945, "Life" Magazine devoted a special edition to the armistice and Japan's surrender. General MacArthur was on the cover.

By February 15, 1942, Singapore and Rangoon had fallen into Japanese hands. It was imperative to defend the road through Burma to China. This would be done by ten Tigers face-to-face with hordes of Japanese Zero Sen fighters. One against thirty, the Tigers would attack behind Japanese lines, bombing all the bridges on the Salween River and stopping the enemy advance in its tracks. Between the months of March and April of that year, Chennault and his Flying Tigers destroyed. more than 300 Japanese aircraft, while suffering a loss of only 15 planes. All America had its eyes fixed on "the cowboys of the sky" whom President Roosevelt was already comparing to the air heroes of the Battle of Britain. It was at this time that Republic Studios filmed "Flying Tigers", with John Wayne in the starring role, that of Woody Jason, inspired by Claire Chennault and Tex Hill.

On July 4, 1942, after very confusing and intricate political maneuvers, the Flying Tigers were disbanded and replaced by the 23rd Fighter Group and the 14th Air Force. Out of patriotism, Chennault accepted to remain in the Army Air Force in China accompanied by a few friends, among whom were Tex Hill and Ed Rector.

In April, 1942, the road to Burma was occupied by the Japanese. The situation was tragic, and the China Air Task Force with Chennault lacked everything. Once again, "old leather face" would save the situation with an idea only a genius could have thought of, while still being undermined by his superiors, Generals Stillwell and Bissel. Also nicknamed by the Chinese "Great virtue, great wisdom and great generosity", Chennault then invented the

The Army Air Forces slogan "Keep 'em Flying", symbolized by B-17 Flying Fortresses in flight. This heavy bomber epitomized American strength.

first air bridge in history. He would have himself supplied over the Himalayan Mountains. A completely crazy and almost suicidal idea, but one which was to work incredibly well. The scenario was unbelievable. Tons of various materials and fuel were shipped by cargo vessel from the U.S. via Africa to Calcutta, India. The 10th Air Force then transported these "treasures" to China, flying at altitudes higher than 8,000 meters through terrible climatic conditions without any pressurization.

This air bridge was to defy all the laws of logic and courage and was nicknamed "Flying the Hump". For good luck and to brave the elements, the pilots and their crews who flew the Hump painted the backs of their leather jackets with Chinese "dolls" and inscriptions. The most famous jacket of this fantastic war episode remains without any doubt the "China Doll" immortalized this year by Avirex. To commemorate the Hump, Avirex has also manufactured another extraordinary painted jacket: "The Bengal Wing Officers' Club".

Chennault's Air Bridge was to work until the end of the war: from 1942 to 1945 the Hump flights would transport into China more than 40,000 tons of material and fuel. All during that time the ex-chief of "The Three Men on a Flying Trapeze" organized the reconquest of China, alleviating the burden of the American Army fighting in the Pacific...

Thanks to the Hump, Chennault received his supplies and could make his airplanes fly. However, the Curtiss P-40s and Flying Tigers looked pretty beaten up: the holes made by machine gun bullets were plugged up with chewing gum; missing parts were replaced by those of shot down P-40s and even some from Zeroes. But, with the energy generated by desperation, the Tigers beat the Japanese and attacked their convoys in the South China Sea, permitting General MacArthur (who always wore a leather A-2 and Raybans) to intensify his offensive attacks in the Central and South Pacific.

In spite of his heroic and desperate action, Chennault had to face various sordid plots to disgrace him. On December 14, 1943, during a secret conference in Cairo, Roosevelt met Chiang Kai-shek and Chennault. "Old leather face" had another crazy idea: to install in China B-29 bases to accelerate a surprise raid of great importance on Japan. But Chennault's enemies were to let the situation decay... This permitted the Japanese to advance undaunted into China. He was ultimately named Commander in Chief and General Stillwell was replaced by General Wedermeyer...

Once again this devil Chennault and his Tigers were to achieve incredible goals, decimating Japanese airplanes while not losing more than a few dozen of their own planes.

As of January, 1945, the Americans crushed the Japanese in the Pacific and Chennault did the same in China. Yet this authentic, 100% American hero was again to be the victim of plots. In July 1945 he was mandatorily retired for health reasons...

That was the moment chosen by Warner Brothers to produce "God is my Co-Pilot", a superb homage to Chennault and the Flying Tigers. In this movie the actors Raymond Massey, Dane Clarke and Andrea King would portray Chennault, Tex Hill, and company.

"But where is Chennault?" asked General MacArthur, indignant at the injustice suffered by his friend during the capitulation ceremony of the Japanese on the decks of the warship "Missouri" in Tokyo Bay in September, 1945.

The old solitary Tiger had left for new adventures! During the Chinese Revolution, he was to organize the retreat of his friend Chiang Kai-shek to Formosa. He would also be behind the creation of the Flying Tigers Airline with his old friend and ex-AVG pilot Bob Prescott. Today, the Flying Tigers Airline is still an important American air cargo company. Chennault also intervened on the sidelines during the Korean War. He died of cancer of the throat on July 27, 1958, while organizing a new AVG, the International Volunteers Group, a private air action group.

The life of Chennault and his Flying Tigers! The most fabulous sagas are tightly intertwined with the heroes, the leather jackets, the P-40 Curtiss airplanes with shark teeth, the cowboys of the skies, and adventure. To finish with the Tigers, we have found letters from pilots which clearly show the importance that their leather jackets had. Here are a few extracts: "I sold my flight jacket to Jack and impatiently await a new delivery" and "when we were in Los Angeles, I bought a flight jacket, cowboy boots and revolvers...", or: "Here a leather jacket costs more than $360 and I am very careful with mine," and lastly: "John wanted to bet me my B-3 at poker. I refused. That good old crusty protected me from the Japs". The letters of heroes in leather... ■

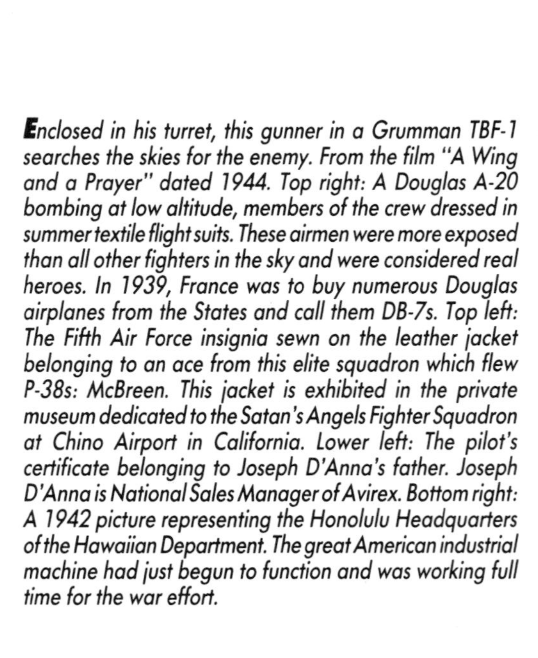

Enclosed in his turret, this gunner in a Grumman TBF-1 searches the skies for the enemy. From the film "A Wing and a Prayer" dated 1944. Top right: A Douglas A-20 bombing at low altitude, members of the crew dressed in summer textile flight suits. These airmen were more exposed than all other fighters in the sky and were considered real heroes. In 1939, France was to buy numerous Douglas airplanes from the States and call them DB-7s. Top left: The Fifth Air Force insignia sewn on the leather jacket belonging to an ace from this elite squadron which flew P-38s: McBreen. This jacket is exhibited in the private museum dedicated to the Satan's Angels Fighter Squadron at Chino Airport in California. Lower left: The pilot's certificate belonging to Joseph D'Anna's father. Joseph D'Anna is National Sales Manager of Avirex. Bottom right: A 1942 picture representing the Honolulu Headquarters of the Hawaiian Department. The great American industrial machine had just begun to function and was working full time for the war effort.

HEADQUARTERS HAWAIIAN DEPARTMENT

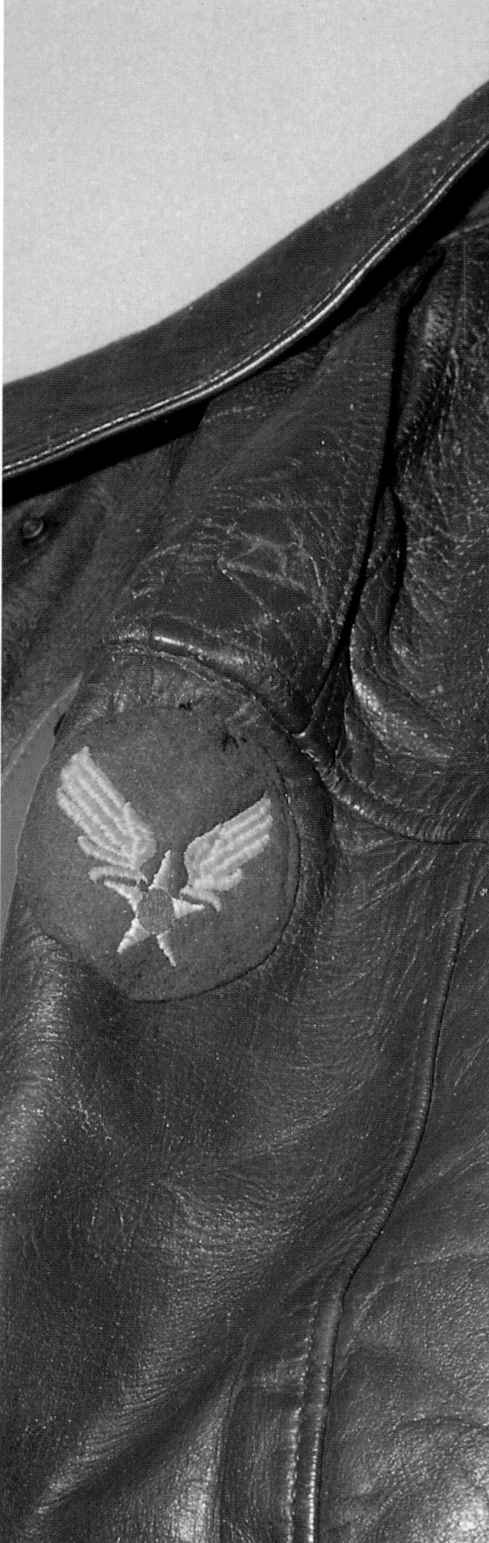

"Tojo's Nitemare". An A-2 flight jacket belonging to a pilot in the Pacific. The flags symbolize the number of Japanese airplanes shot down. This is a very beautiful example which is part of Jeff Clyman's collection. You can see it displayed in the Cockpit store in New York. Upper left: The takeoff of a Grumman F-6F fighter from the aircraft carrier "Lexington". After 1943, this airplane was the most effective naval fighter used against the Japanese. Above right: This photo taken in 1944 shows us a U.S. Navy ace at the command of his Grumman F-6F. We know, thanks to the flags painted on the fuselage, that he has shot down five enemy airplanes. Below it, another 1944 picture taken in the South Pacific: The briefing before departure of an SB-2C "Hell Diver" pilot of a Navy squadron. Upper right: One of the many Naval aviator insignia manufactured by Avirex.

AVIREX LTD.

TOKYO
YOKAHAMA
TOJO'S
NITEMARE

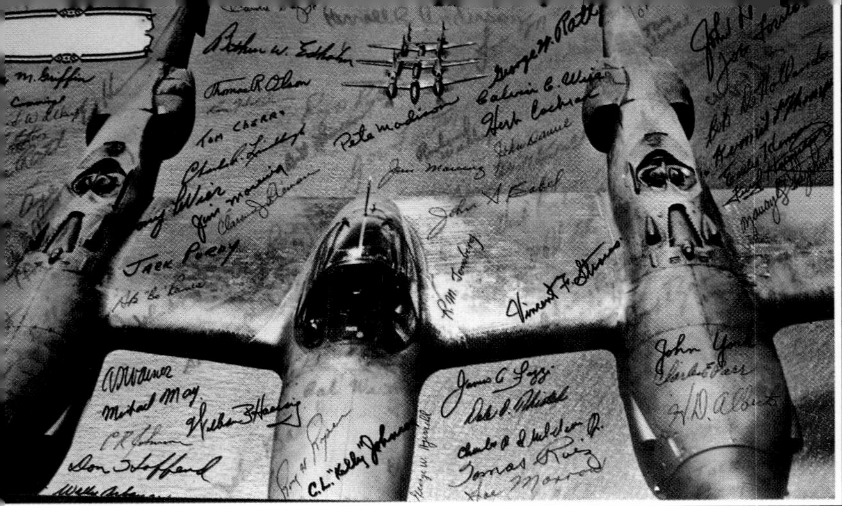

***A** double page dedicated to an elite group made up of aces and heroes: the 475th Fighter Squadron, Satan's Angels, of the 5th Air Force. This original painting, which has become a poster sold at Chino Airport, represents a P-38 in the thick of it in the South Pacific. Satan's Angels had a reputation for being hot pilots and having an almost kamikaze attitude, but these pilots were among the most decorated in the war. Another squadron, the 359th, also fought in New Guinea, and their mascot, in addition to the P-38, was a Gremlin, which was later used in a publicity campaign. Above: A P-38 covered with signatures of surviving Satan's Angels at a reunion of veterans in 1965. As always, the A-2 leather jacket belonging to McBreen in its museum window at Chino Airport in California.*

PUDGY
131

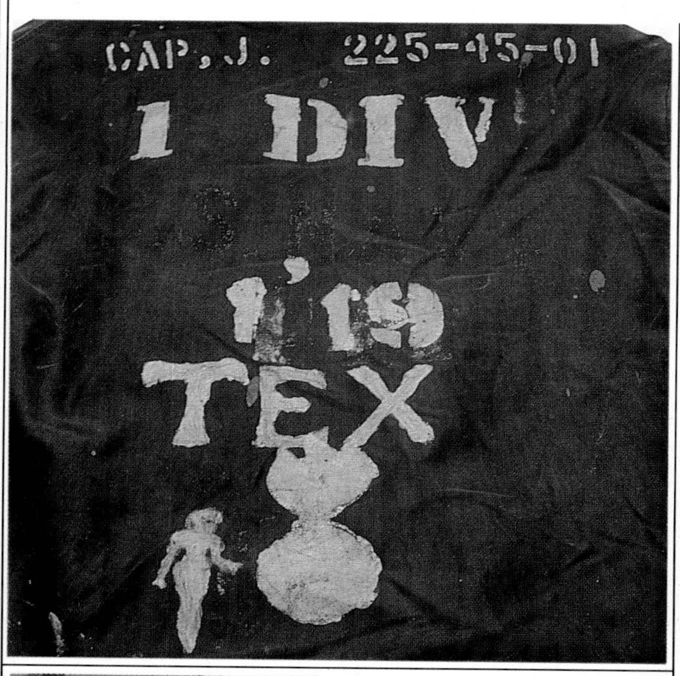

The fabulous nose of a B-29 bomber, the airplane which changed the face of the war in the Pacific, a monster even more powerful than the B-17 Flying Fortress. Above: The back of an Avirex hand-painted jacket and an AAF ad with the slogan "Keep him flying". Opposite: A complete B-24 crew posing for the famous photos symbolizing the 100th mission. They are all wearing their A-2 jackets. An ad from 1942 extolling the merits of a medium bomber: The B-25 Mitchell "Jack of All Raids".

NAVY — MARINES

Chester Nimitz, admiral of the fleet, is top Navy boss in Pacific.	**John Hoover,** vice admiral, succeeded Towers as Nimitz' deputy.	**Forrest Sherman,** rear admiral, is chief of staff to Admiral Nimitz.	**William F. Halsey,** admiral, led Third Fleet in last attacks on Japs.	**Alexander Vandegrift,** general, led at Guadalcanal, is Marine boss.
Raymond A. Spruance, admiral, commanded the U.S. Fifth Fleet.	**Thomas Kinkaid,** vice admiral, led Seventh Fleet at Philippines.	**John S. McCain,** vice admiral, led Task Force 38 for past 14 months.	**Marc Mitscher,** vice admiral, led the hard-hitting Task Force 58.	**Roy S. Geiger,** lieut. general, is the boss of all the Marines in Pacific.
John H. Towers, vice admiral, was Nimitz' deputy for most of war.	**Richmond K. Turner,** admiral, led amphibious forces in the Pacific.	**Fred Sherman,** vice admiral, led task force in major carrier attacks.	**Daniel Barbey,** vice admiral, commands the 7th Amphibious Force.	**Lemuel Shepherd,** major general, commanded 6th Marine Division.
Charles Lockwood, vice admiral, bosses all the Pacific submarines.	**Theodore Wilkinson,** vice admiral, led one of the Leyte landings.	**Harry Hill,** vice admiral, led landings at Tarawa, Eniwetok, Tinian.	**William W. Smith,** vice admiral, is the service officer for the Pacific.	**Holland Smith,** lieut. general, was an early Marine boss in Pacific.

ARMY — AIR FORCES

Douglas MacArthur, general of the Army, is Pacific Army boss.	**Joseph W. Stilwell,** general, was CBI boss, now leads Tenth Army.	**Albert Wedemeyer,** lieut. general, is boss of Army in China Theater.	**Carl Spaatz,** general, went from ETO to lead the Pacific air forces.	**Curtis E. LeMay,** major general, is the chief of staff to Spaatz.
Walter Krueger, general, helped to reconquer Philippine Islands.	**Robert Eichelberger,** lieut. general, led Eighth Army in Philippines.	**Raymond Wheeler,** lieut. general, became India-Burma boss in 1945.	**James H. Doolittle,** lieut. general, moved 8th Air Force to Pacific.	**Nathan Twining,** lieut. general, is in command of 20th Air Force.
Innis P. Swift, major general, commanded I Corps of Sixth Army.	**Charles Hall,** lieut. general, led the XI Corps in the Philippines.	**Oscar Griswold,** lieut. general, led the XIV Corps, freed Manila.	**Claire Chennault,** major general, led 14th Air Force, has just quit.	**George C. Kenney,** general, sold MacArthur on Pacific airpower.
John R. Hodge, lieut. general, was one of the first ashore at Okinawa.	**Franklin Sibert,** major general, led the X Corps in the Philippines.	**Richard Sutherland,** lieut. general, is MacArthur's chief of staff.	**George Stratemeyer,** major general, is the air commander in CBI.	**Charles Stone,** major general, is new leader of the 14th Air Force.
Robert Richardson, lieut. general, bosses Middle Pacific Command.	**Wilhelm Styer,** lieut. general, has post of MacArthur's supply chief.	**Archibald Arnold,** major general, led the 7th Division on Okinawa.	**Ennis Whitehead,** lieut. general, is commander of 5th Air Force.	**Paul Wurtsmith,** major general, is in command of the 13th Air Force.

Opposite: The exact replica of a B-3 bomber jacket, manufactured by Avirex. Notice the patch pocket which was usually worn on the right as well as the leather reinforcements on the sleeves and shoulders. These reinforcements existed more than 40 years ago and were supposed to protect the sheepskin from cuts and nicks. Notice also the belt straps made of a different, more resistant, leather so as not to break under pressure. Above: A patch by Avirex in memory of training operations aboard aircraft carriers. At right: The 45 important commanders of the Pacific war: From MacArthur to Nimitz, not forgetting Sherman, Spaatz, Chennault, Southerland, Le May...

ENOLA
GAY

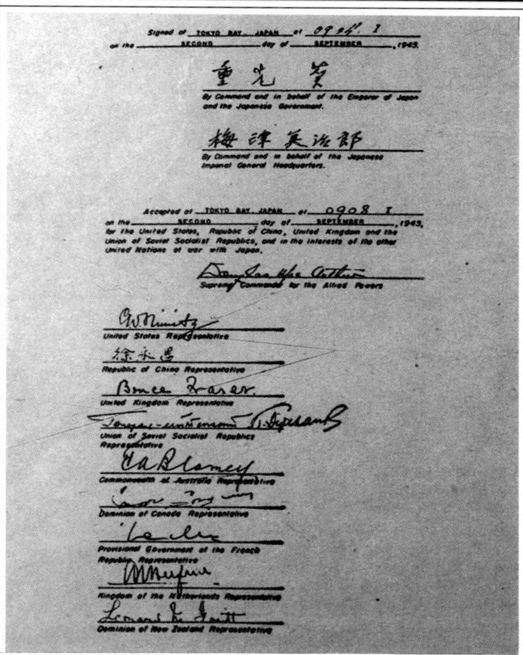

Signed at TOKYO BAY, JAPAN at 0904 I
on the SECOND day of SEPTEMBER, 1945.

By Command and in behalf of the Emperor of Japan and the Japanese Government.

By Command and in behalf of the Japanese Imperial General Headquarters.

Accepted at TOKYO BAY, JAPAN at 0908 I
on the SECOND day of SEPTEMBER, 1945, for the United States, Republic of China, United Kingdom and the Union of Soviet Socialist Republics, and in the interests of the other United Nations at war with Japan.

Supreme Commander for the Allied Powers

United States Representative

Republic of China Representative

United Kingdom Representative

Union of Soviet Socialist Republics Representative

Commonwealth of Australia Representative

Dominion of Canada Representative

Provisional Government of the French Republic Representative

Kingdom of the Netherlands Representative

Dominion of New Zealand Representative

On August 6, 1945, the "Enola Gay", a B-29 named for pilot Paul Tibbet's mother, dropped a 20-kiloton atomic bomb nicknamed Little Boy on Hiroshima. Three days later, a second A-bomb weighing 4.5 tons and carrying the name Fat Man fell in its turn from another B-29 named "Bock's Car" over Nagasaki. Bob Etherly, the pilot of "Straight Flush", one of the B-29 weather recon airplanes which watched the operation, said he would never recover from this double bombing. On Sunday, September 2, 1945 at 9.08 a.m., General MacArthur signed the official document for Japan's surrender. The act took place on the bridge of the "Missouri", anchored in Tokyo Bay. World War II was officially over.

EUROPE

"THE FORTRESSES OF HELL"

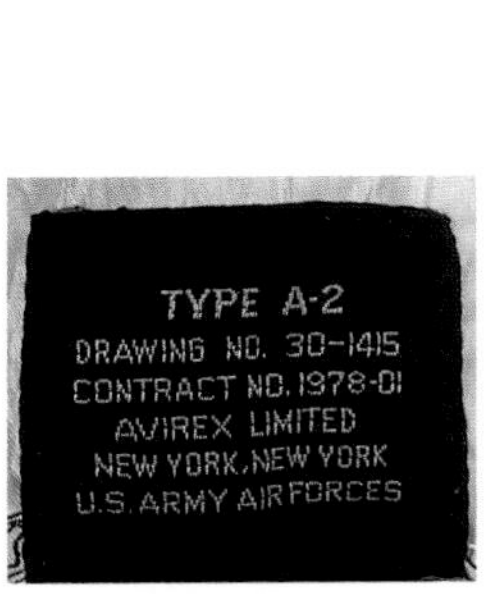

During World War II, the B-17 Flying Fortress symbolized the strength of conquering American air power. This legendary heavy bomber was one of the main aircraft which turned the tide of the war. In May, 1941, seven months before Pearl Harbor, the U.S. furnished the first B-17s to England. In June, 1943, at a time when U.S. industrial power was mobilized for the war, the U.S. sent 8,685 B-17-Gs to the European front. Nicknamed "The Fortress of Hell" the B-17-G was the last word in bombers, armed with 13 heavy machine guns carrying a maximum bomb load of 10,000 1bs. Inside were ten men dressed in B-3 or A-2 leather flying jackets. The B-3, a bomber jacket with such panache that Patton and Montgomery made it part of their favorite uniform. Montgomery immortalized the English version of the pilot's sheepskin jacket, while Patton immortalized the American B-3, but the jackets would both become almost as famous as the heroes themselves after the war.

Ten members of a B-17 Flying Fortress crew photographed in 1944 in England just before a mission. Opposite: A pilot of a P-51 after a mission. He has written on his hand the particulars of his mission, his rendez-vous and his home base. An original picture also taken in 1944. Advertising which appeared in an American magazine in 1943 describing new pilots' equipment with the caption, "Dressed to Kill. In this outfit, our airmen deliver death to the enemy."

In his movie "The Mission", Steven Spielberg tells an incredible adventure of a B-17 Flying Fortress gunner during World War II. Not just any gunner. This gunner is scruntched up in the ball turret under the fuselage of the giant bomber. A pretty dangerous position. Not only was the space cramped, but this gunner's "turret" was only connected by a small hatch to the inside of the airplane. The gunner entered by a trap door and was, therefore, isolated for hours in an almost insufferable fetal position. But this was not all. In case of accident, or when the landing gear failed to extend, the turret was in an exposed vulnerable position and would be completely crushed if the B-17 had to make a "gear-up" belly landing. The "ball turret" gunner was in an unique position because he had to be very small in stature to fit the small space and at the same time physically powerful enough to operate both the twin .50 caliber machine guns and the manual electric tracking mechanism. The ball turret was extremely important because most attacks by enemy fighters were made either from head-on or from below and behind, where only the twin guns of the ball turret could defend. Missions lasted up to 15 hours and the ball turret belly gunners were an exceptional breed since the turret had no heat for the gunner except to keep the sighting glass clear of ice. The gunner was wrapped in his B-3 sheepskin jacket, A-3 sheepskin pants, A-6 shearling boots and B-6 shearling helmet. Temperatures were — 40° F and ball turret gunners were special men. They had little opportunity to bail out since they had only a trap door which exited back into the fuselage of the B-17. In "The Mission", the character, Casey Siemaszko, is the gunner aboard the Flying Fortress B-17 "Friendly Persuasion". The nose of his bomber is painted with a voluptuous pinup wearing a flimsy nightgown riding an enormous red bomb. On the front of this projectile the crew members had added two words: "Hello Adolph". Casy is 19 years old, proud to be an American fighting the hated Nazis and, in spite of the lack of comfort, he feels invulnerable wearing his sheepskin B-3 jacket. His model A-3 pants, also in sheepskin, are worn high above his waist and held up by suspenders and, wearing his baseball cap on his head, he poses at ease in front of the nose art of the most powerful bomber. He does not yet know that in a few hours he is going to live through hell. During a mission over Germany, "Friendly Persuasion" is hit. Its landing gear and the ejection system of the turret are blocked. The Flying Fortress has to make an emergency landing on its belly right on the gunner's turret... "The Mission" tells the story of these last minutes of the crucified gunner in an overwhelming manner. But we are at the movies and all ends well. Once again, Spielberg has excelled, finding fabulous jackets and uniforms, not to speak of B-17s and the atmosphere surrounding that era. More than 50 years after the war, the Flying Fortress and its leather flight jackets have the same appeal to the heart and mind. The B-3 used in the movie is no doubt an Avirex recreation.
As a matter of fact, Hollywood's flirting with the world of aviation, its heroes, the war, and the jackets, took off in 1938 with "Test Pilot", which starred Clark Gable, Myrna Loy and Spencer Tracy. A few years later, Clark Gable would really be in the war in the Eighth Army Air Force in England as a waist gunner on the B-17 "Duchess". The incredible success of "Test Pilot", a movie by Victor Flemming, started the fashion for the "intrepid pilots in their

On July 12, 1943, "The Saturday Evening Post" devoted its cover page to the little-publicized U.S. air war against German U-boats in the North Atlantic. The landing signal officer, nicknamed the "Flagman" or LSO, directed aircraft onto the decks of aircraft carriers.

leather jackets". The studios never let that fashion die.
The source of inspiration was to increase, especially when, in 1939, Europe was set on fire by Hitler's German war machine. The Army had invented leather jackets for the needs of the war and Hollywood would make these items of clothing into symbols through about 60 impressive movies that were propaganda documents designed to bring together the American people. In 1939, the year of the Declaration of War against Germany by England and France, the studios brought out "Wings of the Navy". After that came "Flight Command" (1940),

with Robert Taylor, and "I Wanted Wings" (1941), with Ray Miland and William Holden... After the disaster of Pearl Harbor in 1941, the momentum of war films, especially about aviation, increased. The script writers had an easy time since the real facts were there: all one needed to do was to romanticize the exploits of these 100%, real American heroes to whom, in reality, heroics were a part of daily survival. It was relatively easy to arrive at fantastic adventures in movies: simply by telling the truth —it was even more fantastic than any Hollywood fantasy! Still in 1941, Errol Flynn landed the part of the "lion" in "International Squadron". He shared the limelight with another actor, Ronald Reagan, who wore his flight jacket like a real "cowboy of the skies". At the end of 1941, James Cagney and Dennis Morgan arrived on the silver screen in a splendid movie called "Captains of the Clouds", an unbelievable plug for leather jackets. In 1942, Ronald Reagan and Errol Flynn were again reunited with Nancy Coleman to star in "Desperate Journey", a movie by Warner Brothers about flyers escaping from Nazi Germany. Also that year, the ultimate movie classic "Flying Tigers", with John Wayne and John Carroll, would be shown. The latter would be a successful monument comforting the Hollywood producers in their war actions. The rhythm in the production of these hard-core aviation films would again accelerate and the year 1944 would start with a bang with the documentary history of the B-17 Flying Fortress "Memphis Belle". This time the movie was produced by the U.S. Army Air Forces. The heroes of this movie were the crew and their incredible Flying Fortress, nicknamed in honor of pilot Captain Morgan's fiancée, Miss Margaret Polk, from Memphis, Tennessee. "Memphis Belle", the first bomber on the European front to have completed 25 missions over Germany, without counting a good ten Messerschmitts shot down, was immortalized. Aboard it, ten real heroes wearing leather: Harold Loch, Cecil Scott, Robert Hanson, James Verines and Captain Robert Morgan, Charles Leighton, John Quillan, Casimer Nastal, Vincent Evans and finally, Clarence Winchel. This film, directed by Lieutenant Colonel William Wyler (Hollywood great Billy Wilder at war) and narrated by Captain Edwin Gilbert, was to have a tremendous success and make the B-17 Flying Fortress a memorable and symbolic star.

***A** famous painting representing General Patton wearing his personalized B-3 during the battle of Bastogne. Note the ivory handles on the revolvers and the pit bull. The artist is Michael Gnatek.*

But the best was still to come. From 1945 to 1988, Hollywood continued to bite into the subject of World War II, always keeping in the background leather bomber jackets worn by well-known actors. The postwar years would take off with "The Best Years of our Lives", followed by "The Beginning of the End" and "Fighter Squadron", made in 1948, the producers combining the adventures of the Flying Tigers with the exploits of the fighter aces over Europe in which a new actor was starting his career: Robert Stack. In 1949, Gary Cooper immortalized Navy Lieutenant Jonathan Scott, a pure hero in leather, in the movie "Task Force". In 1949, another giant of the screen, Gregory Peck, would also wear the A-2 flight jacket as General Franck Savage in the immortal classic "12 O'clock High", about the 8th Air Force B-17s in Europe.

Since every war movie in which aviation predominated was a success, Hollywood, thrilled by such luck, continued on its way, maintaining in the public eye the legend of the heroes and their flight jackets. In the fifties and early sixties, Hollywood gave us flying Marine pilots in "Flying Leathernecks", Navy aces in "Flat Top", the story of the A-bomb in "Above and Beyond", "Battle Stations" and "China Doll" and, in 1962, two new actors, Steve McQueen and Robert Wagner, immortalized the B-3 bomber jacket in John Hersey's "The War Lover", about B-17s over Europe. The same phenomenon was observed during the seventies with super productions and unprecedented success in films such as "Catch 22", "Tora, Tora, Tora", and "Midway", all of which epitomized the glory and absurdity of war, all wrapped up in a leather flight jacket. After that we would have to wait for the 1980s to find important aviation movies which would again bring back the myth of the bomber jacket. ■

two U.S. pilots wearing the famous leather of heroes: Colonel Francis Gabreski and Captain Robert Johnson.

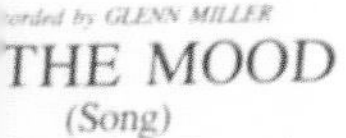

*"**G**len Miller AAF Band"! One of the most beautiful jackets in the Clyman collection. The quality and precision of this painting, done by hand on the back of an A-2 flying jacket, is remarkable. This jacket is enclosed in a display case at the Cockpit. A hand-painted version of "Glen Miller", in homage to the famous musician, composer and American orchestra leader, shot down in his airplane in 1944 during the D-Day landing period. The mystery of his death was never solved. Was he shot down by a German plane... or an English one? Far left: Two of the greatest top-scoring American aces. In color, wearing a B-10 flight jacket, is Colonel Francis Gabreski, who has 31 swastikas on the fuselage of his P-47D, meaning that he had shot down 31 German planes. In black and white, another flying ace in an A-2, Captain Robert Johnson, who claimed 22 victories, also inscribed on his P-47D. Top left: A bombing mission on Ploesti Refinery in Rumania by a B-24 Liberator. Upper Right: An Avirex motif with Clark Gable, who was a Captain in the Army Air Forces and a gunner on the B-17 "Duchess".*

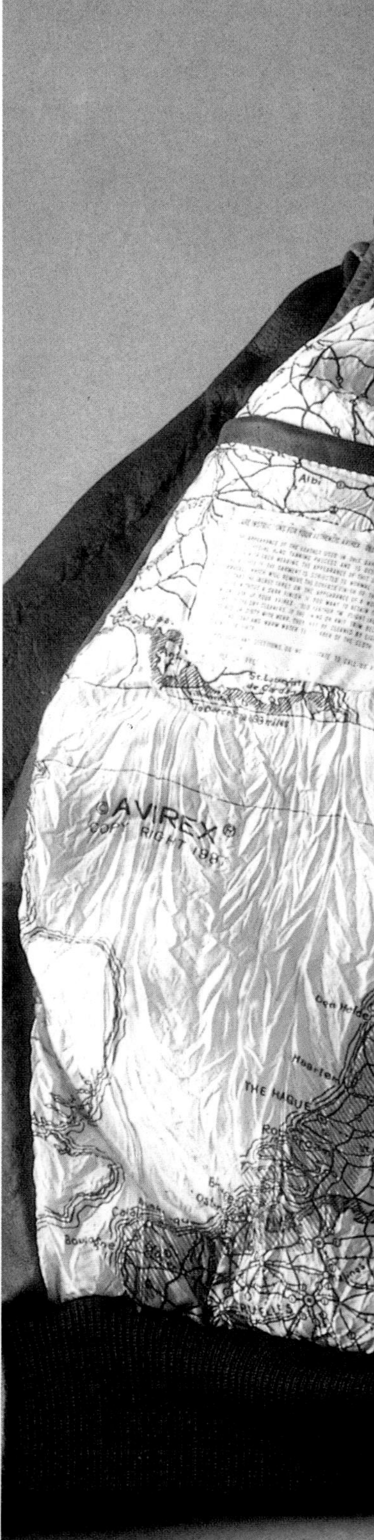

When the pilots parachuted into a hostile country, or when their airplanes were shot down, they customarily had a detailed map of the area and country sewn inside the lining of their flight jacket. To commemorate this custom, Avirex has created a collection of A-2 flight jackets whose lining is the map of that area. Far right: The lining of this A-2 jacket is the escape map of France that American and British pilots used in World War II. Top left: Three members of a B-17 crew; on the back of their jackets, they have painted the name of their bomber, "Pistol Packin' Mama". As always, the painted bombs represent the number of missions and the swastikas symbolize the enemy planes shot down. At right: The "V for victory" used in publicity for these air transport planes. A "Life" cover dated June 6, 1943, of the members of the Eighth Air Forces Bomber Group all sporting their leather A-2 jackets. Top right: The painted back of an Avirex jacket in remembrance of the U.S.O. tours.

a leather A-2 flight jacket lined with a reproduction of a World War II pilots' escape map: Avirex pays homage to these air men.

***The** inside of the cockpit of B-17 Flying Fortress "Piccadilly Lilly", today based in The Chino Airport Air Museum in California. It sits right in front of the visitors' entrance, a giant welcoming old friends. It is a real monster, but the interior space for the 10-man crew is extremely small. Above: A Coca-Cola ad dated 1943 whose source of inspiration was the friendship between the American and RAF pilots. An A-2 flight jacket signed Avirex. Top right: A trio of singers on a U.S.O. tour in England in 1944. This original picture was sent to Jeff Clyman by one of the singers. Bottom right: One of our favorite pictures, not only because of the angle at which it was taken, but because of the pilots' facial expressions as well as the sheepskin B-6, B-9, and D-1 jackets, and the cloth M-1941 jacket they were wearing. An original picture taken with a P-47 of the 15th Air Force, operating in the Mediterranean area, in the background.*

The B-17 Flying Fortress "Piccadilly Lilly" welcomes visitors to The Chino Airport Air Museum in California.

Avirex helped in the return to th
"Thunderbird", symbol of America

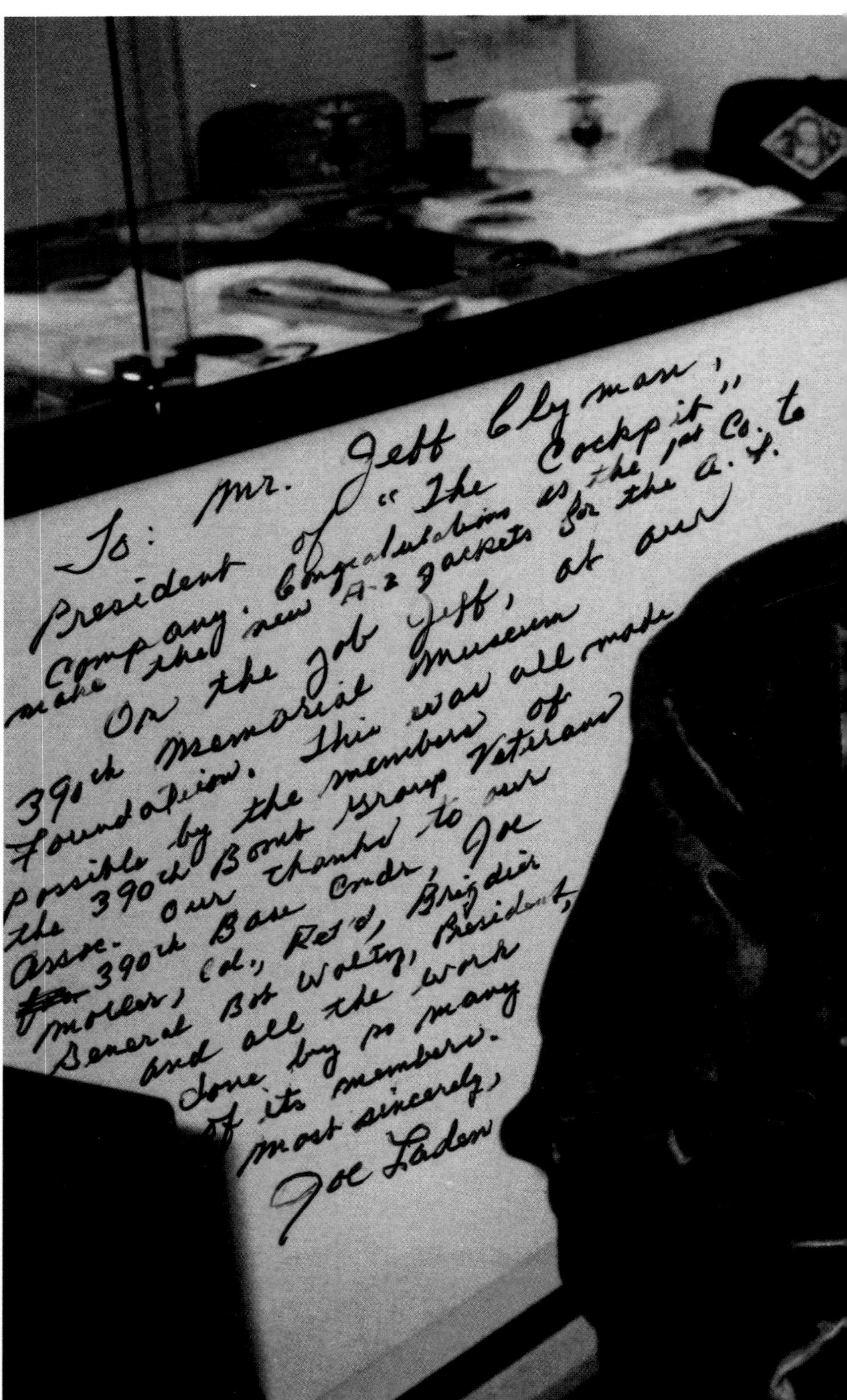

This double page is dedicated to the "Thunderbird" B-17 Flying Fortress and to its original crew member Joe Laden of the 390th Bomber Squadron, which operated in Europe between 1943 and 1945. The bomber was damaged and used after the war for photo-mapping Europe. In 1987, Avirex sponsored the return of the "Thunderbird" to the U.S. Top right: The 1988 arrival of the "Thunderbird" in New York, above the Statue of Liberty and escorted by Mustang P-51s. Centre right: J.K. West, the pilot who flew the "Thunderbird" across the Atlantic in 1987. Bottom right: This 30-meter-long mural painted by famous aviation artist Keith Ferris showing the "Thunderbird" in action is exhibited in The National Air and Space Museum in Washington. At right: Joe Laden sent this autographed picture to Jeff complimenting him on the quality of the jackets manufactured by Avirex. The veteran pilot wears his old A-2 flying jacket adorned with the 390th Bomb Squadron patch. Thanks Joe! Above: The back of an Avirex jacket painted by hand in honor of Glen Miller, with an original picture, taken in 1944, of the real Hell's Angels in front of the painted nose of their B-17. Each member of the squadron signed his name on the fuselage of the Fortress. The meeting of A-2s and B-3s. Legend has it that the first motorcycle gangs in California, founded by ex-military men, were inspired by these flying aces.

.S. of the B-17
night.

The painter, Jim Dietz, is inspired by the leather jackets of World War II heroes.

This superb surrealistic still life is the work of an American painter, Jim Dietz, to whom we also owe the marvelous painting on our back cover. Dietz named his work "The Few" after the few British fighter pilots who fought and won against the German Nazi hordes in the Battle of Britain in 1940. Note the quality of the sheepskin jacket, an English Irvin, and the aviation atmosphere which is recreated. A true work of art! The magic of military jackets and the nostalgia which emanates from them have always inspired Dietz, who is, in fact, only about 40. Above: On the cover of this copy of "Life" Magazine, dated May 15, 1944, General Montgomery wears his inseparable Irvin in sheepskin. Below it: A unique example of a German Luftwaffe pilot's jacket, with all its "hated decorations" in perfect shape. Far left: Two enemy fighter pilots, the first wearing a leather jacket similar to the foregoing one and the second a cloth jacket with a knit collar.

An original document from 1944. The two pilots wearing A-2 flight jackets are showing the German cities they have bombed. The photograph was captioned: "After the mission" by the Army Air Force authorities. Top right: The back of an Avirex jacket painted by hand in memory of the pilots on the European front. The original motif is taken from a painting on a B-17's nose. Bottom right: A group of B-17s during a bombing mission above Germany: "Somewhere between heaven and hell." Above: Briefing Eighth Air Force fighter pilots in Europe. A rare color photo taken in 1944. They are all wearing A-2s or B-3s. A Bendix advert in 1944 in which P-40s and A-20s are flying over tanks. "The invisible crew". Victory on its way!

for the pilots, the leather jackets were "second skins", good luck clothes to cheat death.

On the 7th of December, 1941, the Japanese pulled a surprise attack on Pearl Harbor and wiped out a part of the American Naval Fleet as well as numerous airplanes, among them Curtiss P-40s and B-17 bombers which had just arrived. This disaster incited the United States to start an intensive program of airplane production. From the beginning of 1942, America produced 47,836 warplanes, among which were 10,769 fighters and 12,627 bombers. In 1943, the U.S. had 85,898 airplanes on all fronts. This record was beaten in 1944 with a total of 96,318. Today's aviation specialists, collectors, and buffs have nicknamed these old, restored combat airplaines "war birds". We have decided to show you the most prestigious and well-known of these in action during the war years 1939-45 as well as during current air shows. Far right: Self-portrait of Jeffrey Clyman, taken by an automatic long-distance camera, piloting his own AT-6-D, "Double Trouble". The "T-6 Texan" started in 1938 as a BC-1A Basic Combat Aircraft. In 1942, it became the best known and most famous of trainers. America was to produce more than 15,000 of these trainers, which were still in use twenty years after the war. Bottom right: "The Six of Diamonds", the T-6 flight team sponsored by Avirex and the Cockpit catalog. Top right: A T-6 Texan taxiing during the Valiant Air Command show at Titusville, Florida, in March, 1988. Centre right: A 1942 original photograph of a T-6 zooming through the skies.

WAR BIRDS

WAR BIRDS

Justly nicknamed the Rolls-Royce of the sky, the North American P-51 Mustang was certainly the fastest fighter, as well as the most efficient and radical. Appearing in the skies during the second half of the war, the Mustang is supposed to have been the airplane that ultimately gave supremacy to the Allies. It was the product of two aeronautical technologies: that of the U.S. and Great Britain, which equipped it with a fabulous piston engine signed Rolls-Royce Merlin. In 1942, the U.S. produced the Mustang with a license-built Rolls-Royce Packard engine and named it the P-51-B. In 1944, it became the Mustang P-51-D, an even better version. Another version existed, the P-51-H, which was capable of flying at 487 miles-per-hour at 30,000 feet. Today, Mustang P-51s are the incontestable show-stoppers at all air shows and are sought after by all collectors. Far left: "Six Shooter", a Mustang P-51 photographed during the Sun and Fun air show at Lakeland, Florida, in April, 1988. Top left: The lineup of Mustangs and the two P-51s in formation were also shot in Florida. The last picture, below, is an original Air Force shot dating from 1944. The Mustang and its pilot were immortalized on the European front by an Army photographer.

WAR BIRDS

On August 19, 1940, the Americans produced a twin-engine medium bomber which was to become the illustrious North American B-25 Mitchell. It was so named in honor of General "Billy" Mitchell who, since the early 1920s, had insisted that the U.S. start building up its airforces. Up until 1944, North America was to manufacture increasingly sophisticated versions of the B-25, with the subsequent models A, B, C, D, and especially the H and J. This was an exceptional plane, equipped with heavy artillery which included a 75-mm cannon in its nose. But the symbol of American strength remained the legendary B-17 Flying Fortress, a heavy bomber with four motors and an exceptional bombing capacity. The prototype of the B-17 appeared on July 28, 1935, and was continually improved upon from then on. Then came the B-17-B, and finally the G, of which 8,685 would be built, mostly for combat on the European front. But the last word in heavy super-bombers was to remain the giant Boeing B-29 Super Fortress. On December 18, 1942, the B-29 prototypes left the factory. They were manufactured by Boeing, Bell Aircraft Company and the Glen L. Martin Co. The first B-29s became operational on the Indian and Chinese fronts in the spring of 1944, where Jeff Clyman's father flew these amazing aircraft. In all, 3,970 B-29s were built. The most famous B-29s are the "Enola Gay" and "Bock's Car", which dropped the first atomic bombs on Hiroshima and Nagasaki. The B-29-45 MO ("Enola Gay" model) could fly at an altitude of 31,850 feet at a speed of 358 mph. With an amazing bomb capacity and defensive fire power, they feared no foe. Far left and top left: These photos of B-29s and a B-17 are original documents. At left: The pilots briefing is taken from the movie "The Wild Blue Yonder", and the B-25 in flight was photographed at Harlingen, Texas during an air show.

If we have chosen to show you four other American fighters and bombers: the P-38, the P-40, the Corsair, and the B-26, by pictures taken from advertising copy, it is not for lack of real photographs, but to better describe the American mentality of the period. The aim was to mobilize the entire American people. Huge advertising campaigns appeared in widely distributed magazines such as "Life" and "The Saturday Evening Post". It was very important to inform the American people and to make them cooperative, responsible and sensitive to the war effort. The Allison and Cadillac Motor Companies projected the image of the famous P-40s, which had distinguished themselves in China and in the Pacific arena. But the Army Air Forces themselves (AAF) strongly encouraged patriotism, and the themes of unity and courage. From 1941 to 1945, their advertising campaigns obtained exceptional results. The Lockheed P-38 Lightning, put into production from 1940 to 1945, 9,923 in all being produced, was the most original and the most controversial of all warplanes. With its three fuselages and twin Allison engines, the P-38 maintained the record for the most Japanese airplanes downed. It was used as a fighter and light bomber for both day and night, and as a reconnaissance plane, etc. The notoriety of the Curtiss P-40 is hard to beat, and the American Volunteer Group, later Chennault's Flying Tigers, made it famous. The Vought F-4-U, also called the Corsair, is the fighter which gave air supremacy to the U.S. in the South Pacific. In certain respects, the specialists affirm, it was even superior to the Mustang P-51. It was manufactured for more than ten years after the war, and was used in combat until 1969. In all, 12,681 Corsairs were built. The Martin B-26 Marauder, a medium bomber, was the equivalent of, and performed as well as, the B-25 Mitchell. It was present on all fronts from 1942 to 1945.

Here is positive evidence of the Allison engine's ability to survive an Axis dogfight — to take a hail of lead and keep flying. Punctured in seventeen places, this engine carried its plane and pilot back to safety at an R. A. F. air base in Libya. This engine, later returned to this country and now in the "Arms For Victory" Exhibit of General Motors in Detroit, gives dramatic proof that when the Nazis shoot up an Allison they can't count on shooting it down.

Curtiss P-40 (U. S.) The British call it the Tomahawk or the Kittihawk

Qualitative superiority counts! From Africa, the Middle East, the South Pacific, Russia, the communiques report that nothing in the Axis air armada can match the sharp-nosed fleetness of this liquid-cooled engine. It's a matter of record that predates Pearl Harbor, how Allison-powered planes can dish it out.

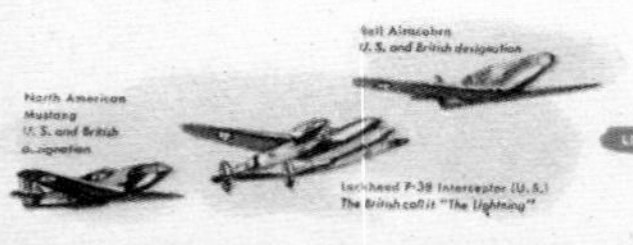

In the vanguard of the fighter craft t
the U-boat down are fighter planes
by the Allison—America's first and
liquid-cooled aircraft engine.
To their splendid list of achieve
combat, these stalkers of hidden devil
added a gratifying record
for dependability—thus
demonstrating not only
the skill and daring of
the combat pilots who fly
them, but also the inher-

CADILLAC MOTOR CAR DIV

W A R   B I R D S

of Hidden Devilfish

edness of the planes themselves. g the vital inner-assemblies for the ngine has been a Cadillac manufac- st since March of 1939. During the that have intervened, *millions* of rts—crankshafts, camshafts, connecting rods, super-charger gears, impellers, and other component units—have "gone to war" bearing the imprint of Cadillac's precision workmanship.

and the ge, pro- ned with Hydra-

Developed over a period of 40 years of building to the principle, "Craftsmanship a Creed—Accuracy a Law," this mastery of exacting manufacture in volume production is but a wartime projection of the peacetime skill that has so long made Cadillac-built products "The Standard of The World."

OF THE AIR—NBC Network

NERAL MOTORS CORPORATION

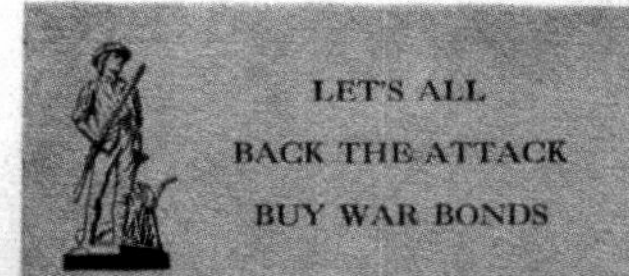

"LET'S GO!"

Says Major Carl W. Payne, of Columbus, Ohio—A.A.F. fighter pilot

Vhen I was home on leave recently, I met a lot of ung fellows who had their hearts set on wearing A.F. wings. Naturally, they fired a barrage of ques- ns my way. And one of the questions I heard most ten was:

Vhat is the best position on the Army Air Forces team . . . 'ot, navigator, bombardier or gunner?

That's a tough one to answer.

was classified as a fighter pilot, myself. And it's a eat job! But on this fighting team of ours . . . as on football team . . . *every* job is equally important.

t works about like this:

The single-seater pilot runs interference.

The bomber pilot takes his crew over to make the uchdown.

The navigator quarterbacks a bomber team. He plots e course from home base to the target—usually ver hundreds of miles of enemy territory—and hits e objective squarely on the nose, with split-second ming.

Then the bombardier takes charge. He draws a bead n a Nazi war plant or a Jap defense line—and sends om one to six tons of destruction screaming down wipe it off the map.

The gunners are the wing-men of the team. And ey're the boys who are seeing some of the rip-roar- g-est, two-fisted-est action of this man's war. Their b is to fight off enemy intercepters . . . and these "Deadeye Dicks" of the air have run up some amazing scores. I guess it's natural for an American to love guns. And maybe that's why A.A.F. aerial gunners are the world's best!

"So you see—I don't know what A.A.F. job you'll like best. *It depends on what you find yourself best fitted for.* But I do know this:

"If it's action you want . . . if it's red-blooded adventure and a chance to fight where you'll do the most good . . .

"If you want the finest training any flying man ever had . . . if you're looking ahead to a career in the air after we've won this war . . .

"Then I say—find out today if you can qualify to join the Air Corps Enlisted Reserve . . . with an opportunity to win your wings as navigator, bombardier, pilot or gunner in the A.A.F.—the greatest team in the world!

"So—if you're seventeen . . . let's go!"

U. S. ARMY RECRUITING SERVICE

A MAJOR AT 22

CARL PAYNE has fl 227 missions, from African invasion to blitz on Berlin. He been decorated 30 ti wears the Di guished Flying Cross Bronze Oak Leaf Cl and the Air Medal wi Oak Leaf Clusters. M PAYNE's score agains Luftwaffe is five and half enemy planes troyed, one prob destroyed and seven aged. Yet neither he his plane has ever scratched!

GET FIT AND STAY FIT!

The future of America depends upon the fitness of the youth of the nation. The AAF urges every man approaching military age to prepare himself physically and mentally for his possible contribution to the defeat of our enemies, and his responsibilities in the postwar world.

Every young man who wants to fly can today find exceptional opportunities for improving himself. Those in high school or preparatory school can take the recommended courses of the High School Victory Corps, which includes physical training, mathematics, physics, and pre-aeronautics. All young m can qualify may become CAP in the nation-wide organiza the Civil Air Patrol, an auxil the Army Air Forces. CAP work with civilian aviation and enthusiasts who give the practical knowledge of plar their construction, and pre-aviation training in su jects as navigation, meteo aerial photography, radio c nications and physical fitne

Don't neglect any opportu prepare yourself in mind a for better service to your your community and yours

KEEP 'EM FLYING!

For information on Naval Aviation Cadet Tr apply at nearest Office of Naval Officer Procur This advertisement has the approval of th Army Navy Personnel Board.

FLY AND FIGHT WITH THE

GREATEST TEAM IN THE WO

AMAZONS

American women played a very important role during World War II. It was no longer a question of them being mere sex objects, such as the pinups which adorned the noses of the bombers or the backs of the jackets, but rather of them taking on such bigger-than-life jobs as pilot, nurse, driver, and movie star, as well as worker in war production factories. The most daring feats of women during the war were accomplished by the Women's Air-force Service Pilots (WASPs), who trained in Texan AT-6s, or piloted medium or heavy bombers such as the B-25 Mitchell and the B-17 Flying Fortress. These women truly reached the stature of amazons. They were especially trained in Texas at Avenger Field Air Base, near Sweetwater, and belonged to the Women's Flying Training Group under the command of Major General Barton Yount. But the initiative to send these young and pretty American women into combat zones, mostly on the European front, stemmed from a truly great woman aviatrix, Ms Jacqueline Cochran, a female luminary in the annals of American aviation.

During the first years of the war, these women wore the same flight jackets as their male contemporaries: A-2s and B-3s. By 1943, the laboratories of the AAF had perfected special uniforms for nurses and WASPs.

During the war, the United Service Organization (USO) organized numerous tours featuring singers, actors and actresses, who visited the fronts as a morale booster. Often, they too would wear the famous flight jackets worn by their contemporaries. This tradition of service would be followed during the Korean and Vietnamese wars, where such stars as Marilyn Monroe and Racquel Welch did not hesitate to expose themselves to potential danger to entertain our fighting men.

The seven crew members (actually Flying Nurses) of a transport based in England in 1942. At this time no clothing outfit had been specified for the women, and in this picture they all wear large B-3 sheepskin bomber jackets. Insets: These two paintings dating from 1942 symbolize the importance of American women during the war. On the one hand, the woman on the home front hoisting the star-spangled banner, and on the other, the woman pilot going into combat.

Jacqueline Cochran was a beautiful young American blonde in love with aviation. It was she who first had the idea of opening a school on a military base for women pilots: Women's Air Force Service Pilots (WASPs) connected to the Women's Auxiliary Ferry Service (WAFS). In four years Jacqueline Cochran, a pilot known for breaking speed records, taught more than 3,000 licensed women pilots, who were to go to war on the European front. The American press at the time had dubbed these pilots "The courageous women".

In 1944, Universal Studios produced a film "Ladies Courageous" nicknaming them "Amazons of the sky". Loretta Young, Geraldine Fitzgerald and Ann Gwyne portrayed beautifully these warriors of the sky in World War II. "Ladies Courageous" was entirely filmed in Long Beach, California, its Army air base supposedly representing Avenger Field in Texas, where the real "WASPs" trained. In 1941, Jacqueline Cochran's 25 best students were sent to England to help the RAF in what were called the "ferry flights".

These women took on the name "Women's Auxiliary Ferrying Squadron". They were so extraordinary that Ms. Cochran then had the idea of creating a special base for her protégées. That is how the Texan saga of the WASPs started, at Avenger Field. The training program was commanded and perfected by Major General Barton K. Yount and by the Chief of the Army Airforces, General "Hap" Arnold. Jacqueline Cochran was also helped by Major Barton "Barney" Giles, sent directly by Washington. To be accepted at Avenger Field, these women had to be between 21 and 24 years old, and to have their pilot's certificate, showing at least 35 hours on light airplanes, a university diploma, and be in top physical shape. Immediately, the instructors noticed that the women learned much faster than the men during training on flight simulators. In contrast, on the purely mechanical side the men were more advanced. The training at Avenger Field lasted 23 weeks. The students worked every day from six in the morning to ten at night. Many were widows of AAF pilots who had been killed in action, while some were single women who just wanted to fight for their country. They wore as a good luck patch the WASP mascot Fifinella, a wasp with a young woman's head designed by Walt Disney, who had created, several years earlier, the Flying Tiger, the famous insignia of the Flying Tigers. As a matter of fact, at the time the Disney studios worked day and night for the Army Airforce and the Navy.

Let us cite among the better-known WASP pilots Rebecca Edwards, Lorena Daly, Anne Armstrong McClellan, Shirley Slade, Florence Knight, Nancy Nesbit, and Jean Pearson. One of the best instructors under Ms. Cochran was the blonde and pretty Helen Duffy.

At the start of hostilities, these women pilots, who trained generally on Texan AT-6s, wore the same uniform as the men. But the smallest of the flight suits and the narrowest of the A-2 leather jackets or sheepskin B-3 jackets were too large for them. As far as clothing was concerned, the "flying nurses" had the same problem as the WASPs. As of 1943, the R and D

A young and beautiful pilot in a picture dated 1941. One of the first World War II women pilots wearing with charm an A-2 leather flight jacket.

Departments of the Army went to work on uniforms for these flying women.

The first jacket perfected for the nurses was the F-1 Type in navy blue wool, standardized on June 7, 1943. This jacket was worn with A-1 Type slacks, also in blue. At the beginning of 1944, they became olive green. In 1944, all the nurses were issued with a new and practical uniform which fitted better: the B-17 Type jacket and the A-13 slacks. This outfit was similar to the one worn by the men: the B-15 jacket and the A-11 slacks. The B-17 jacket would remain a classic with its cuffs and large

waistband in knit, and its fur collar. It was perfectly cut and made the nurses silhouette a lot more feminine.
At the same time, the WASP flight uniform was inspired by the nurses' uniform: the Type B-16 jacket again with a fur collar and the A-12 slacks with four large patch pockets, two on the hips and two on the calves, but also tighter at

Kelly McGillis in "Top Gun", the aviation film of the eighties. She wears a black leather Spencer of the type manufactured by Avirex, which supplied actual Mil Spec jackets for this movie.

the ankles. The shoes worn by the WASPs and nurses were the same: short leather boots with zippers which were very comfortable and had the AAF logo on the side. The soles of the heavy "high" shoes were made of indestructable rubber. Yet many of these women pilots preferred wearing the men's good old leather jackets, thus feeling more appropriately dressed for war. We have also found numerous WASP documents showing heavy Type B-3 or B-6 bomber jackets in sheepskin. The majority of these women flew on P-40, P-47 and P-51 fighters and Martin B-26, B-25 Mitchell bombers and B-17 Flying Fortresses. For all these female fighters, the uniform had the same magic as for their male counterparts, especially where the flight jackets were concerned. However, the WASPs' equivalent to the every-day officers' uniform was a very special pants and jacket, the color of which was called "Santiago Blue". This nickname stuck with the WASPs throughout the war. We should point out that it was not uncommon for war widows who became WASPs to wear their husbands' jackets for sentimental reasons or simply to beat the odds. At Avenger Field all the student pilots wore the cloth flight suit (summer or winter) which they themselves recut, making them a lot sexier.
Many also wore baseball caps with the rim pushed up, thus showing their tomboy side. It was at this time that the major American advertising companies, which worked closely with the Army, began to build their rallying and sensitizing campaign, in which the American woman played the star role.
For many years, the important magazines of the times such as "Life", "Esquire" and "The Saturday Evening Post" would run ads in color which are today considered as definitive examples of that type. On the one hand, the American woman at home, taking care of the children and backing up a husband fighting on the front: a symbol of the stars and stripes showing confidence and comfort. On the other hand, the fighting woman, always very sexy, either healing the soldiers (nurses) or fighting next to the men (WASPs): a symbol of victorious America.
The greatest illustrators of the time, such as Vargas and Norman Rockwell, were recruited for this gigantic publicity campaign, signing calendars and front covers of unforgettable magazines. The movie "Ladies Courageous" was part of this movement to promote the image of the American woman. One of the most beautiful scenes in "Ladies Courageous" remains the one where we see the fighters of the sky at the helm of Curtiss P-40s, Mustang P-51s and P-47s, escorting the B-24 bomber "The Duchess".
But the most heroic women were the nurses, those who fought on the European front, some even flying and dying on missions over Germany aboard such giants as the B-17 Flying Fortress. Nothing was more difficult than to live aboard a bomber: each square inch was accounted for, and the crew would bang into pieces of steel. In addition, the cold and the noise were dreadful, not to speak of the flak and attacks by enemy fighters. For anyone to fly in such conditions required exceptional courage and nerves of steel... It is interesting to note that these unusual women who braved death on each mission were even more superstitious than the men (who took their rabbits' feet with them) and never took off without their good luck doll or teddy bear... Keep 'em flying! ■

LIFE
368
AIR FORCE
PILOT
JULY 19, 1943

On July 19, 1943, "Life" Magazine dedicated a lead article to these women pilots and featured on the cover Shirley Slade, one of the top students at Avenger Field Air Base in Texas. The press finally interested itself in the important role of these heroic women. At left: A photo of a WASP pilot taken in England in 1944 on the European front. She wears one of the new Type A-9 flight suits in cloth (summer model). Below: Women as well as men were involved in experimental flight-suit development.

Centerfold: *The cachet and charm of the crew of a B-17 Flying Fortress based in England appropriately named "Pistol Packing Mama". The three women are wearing A-2 leather jackets, and the second one from the left wears a B-6 sheepskin jacket. This photo was shot for home-front public relations purposes. This same B-17 was also piloted by men whose leather jackets were painted with the name of their bomber. A wartime Coca-Cola ad, dated 1942. Lower right: A student at Avenger Field Air Base training on a flight simulator. Upper right: A woman pilot wearing the flight uniform issued in 1944: a Type B-16 jacket and Model A-12 pants. This pilot's task was to ferry AT-6 advanced trainers from the factory to training bases in the U.S.*

JOAN BENNETT in her American Women's Voluntary Services uniform
Starring in Edw. Small's United Artists Production "Twin Beds"
His Cigarette and Mine
It's CHESTERFIELD
Yours too for a full share of Mildness Better Taste and Cooler Smoking...that's what you and all other cigarette smokers are looking for... and you get it in Chesterfield's Right Combination of the world's best cigarette tobaccos.
Make your next pack Chesterfields... regardless of price there is no better cigarette made today.
EVERYWHERE YOU GO They Satisfy
Chesterfield
200
201
Chesterfield

Marilyn Monroe visiting soldiers in Korea. At the time, this photograph was on the front pages of all American papers. Even in this B-15 flight jacket, Marilyn was irresistible. Lower right: A scene from the film "Bombardier" (RKO, 1943) with Randolph Scott, Ann Shirley and Pat O'Brien. Below: A still shot from "Test Pilot" (Metro-Goldwyn-Mayer, 1938) with Myrna Loy, Clark Gable and Spencer Tracy. From 1938 on, Hollywood would use its stars to mobilize public opinion in America. To commemorate the patriotic action of the film studios, Avirex has manufactured a jacket with a hand-painted "Hollywood at War" theme. At left: An advertisement for Chesterfield cigarettes in a 1942 "Life" Magazine.

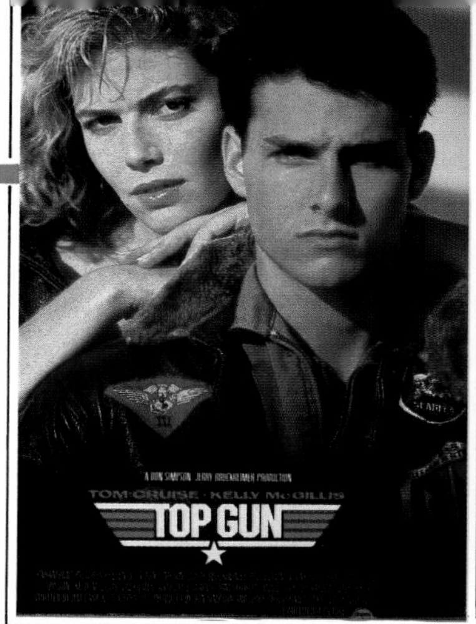

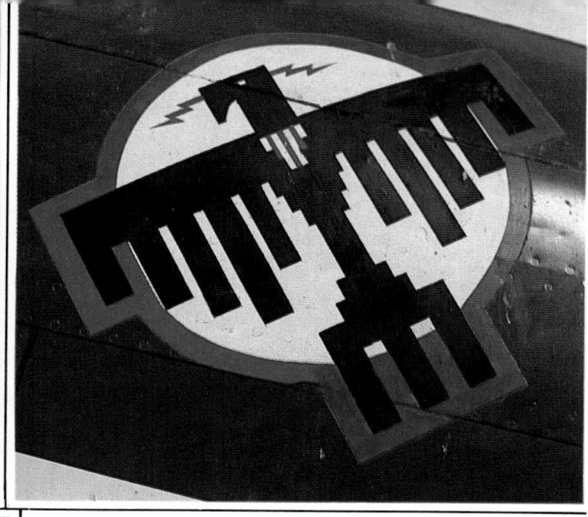

Photos taken at Chino Airport in California, one of the most important "living" American "war bird" museums. Sally, pictured at left, works in the museum and flies a Mustang P-51 with her fiancé. She has on her lap an A-2 leather jacket formerly belonging to one of the "Satan's Angels" pilots, a squadron flying Lockheed P-38s which became famous in the Pacific. The bomber with its shark teeth is a North American B-25 H which has been completely and perfectly restored. Above: The "Top Gun" movie poster, with Kelly McGillis and Tom Cruise. An important page in Jet Age history. Below: A classic pinup of the 1940s.

THE RIGHT STUFF

"THE JET AGE"

Aviation buffs have nicknamed the postwar period from 1946 to 1960 "the Jet Age". Among the hundreds of pilots who regularly challenged the skies, attempting to go ever faster, two men, both pilots in World War II, were to rank among the finest. Their secret? They were said to have "the right stuff", a phrase made popular by Thomas Wolfe in his book about that era. One of these is Herbert "Fireball" Fisher. Nicknamed "Fireball" because of his bright red baseball cap worn on test flights, and receiving almost no publicity, Fisher was chief test pilot for the Curtiss Wright Corp., creator of the P-40. He participated in the invention of the giant, curved propellers to try to beat all aircraft speed records, and would himself fly with such a propeller on his experimental P-47 Thunderbolt in 1946. Fisher was one of the unsung heros of experimental test flying in those years. The other was Chuck Yeager, who, on October 27, 1947, broke the sound barrier piloting the Bell X-1 Rocket Plane. The Korean and Vietnamese wars and finally the "Top Gun" generation would see "antique" jackets, the A-2 and the B-3, replaced by sophisticated flight suits. But even into the sixties there was a hard core of pilots who would always end up flying in leather jackets. The reaffirmation of this mystique took form when the United States Airforce decided in 1987 to officially return to the "good old leather" A-2 flying jackets. Avirex officially delivered the first of these on August 15, 1987.

A fantastic picture by George Hall taken with the real "Top Gun" Fighter Weapons School Squadron in the California skies. The Cockpit shows it as an F-5 Aggressor aircraft, and the pursuing aircraft as an F-14 Tomcat. Upper right: Herbert Fisher in front of his P-47 Thunderbolt with its giant modified, curved propeller, designed to break propeller speed records. Note that no propeller aircraft achieved this result. Lower right: Chuck Yeager in front of a beautiful 1983 flying restoration of his P-51 Mustang, "Glamourous Glennis", named after his wife. Inset above: The label from the Right Stuff collection created by Avirex for an A.C. Delco promotion featuring General Yeager as spokesman.

WEAPONS SCHOOL

The end of the Second World War in 1945 ushered in a new era for aviation clothing. The years of "the Jet Age", as well as the Korean and Vietnamese wars, would see another war, fought out between leather flight jackets and nylon and cloth flight jackets, a war between technology and tradition. But for once, progress and technology would not have the last word. After years of conflict the old leather jacket, symbol of the pioneer spirit of aviation and "second skin" of the fly-boy fraternity, would win what has been called "the great flight jacket war..."

In 1943, the last contracts were issued to suppliers of the A-2 and B-3 flight jackets. Yet the supplies of these fabulous garments were so enormous that pilots continued wearing them until 1945. The B-3 would have an even longer career, since at government insistence so many different suppliers had amassed huge quantities of these unissued sheepskin jackets that they were either issued to aircrews in cold areas or given to rebuild Europe as part of American aid during the Marshall Plan.

1943 is a banner year in the history of flight jackets. It was in that year that the B-10 in olive green cotton canvas cloth, lined with alpaca wool and with a fur collar, was first issued. Its design was comparable to the Navy G-1 jacket. The B-10's career would only last a year. The hard-core pilots are still as attached to the good old A-2 and B-3 jackets as they are to their own girl friends, wives, and children...

Towards the end of 1943 and the beginning of 1944, the R and D laboratories tried to end this sentimental love affair with leather by creating the Type B-15 jacket made of a mixture of canvas and gabardine with a lining in alpaca wool. But, inspite of its thick "mouton" fur collar, the B-15 was neither warm enough nor airtight enough, nor did it have the cachet of the old leather jackets. Until 1947 they would insist on wearing their legendary A-2 and B-3 jackets as if nothing new was happening on the flying jacket front...

In September, 1947, the Army Air Forces were disbanded and replaced by the United States Air Force, which became a separate and independent branch of the American Armed Forces equal to the U.S. Navy or the Army. This date is a milestone, since at the same time propeller-driven airplanes were being replaced by jets. The newly-formed U.S. Air Force took decisions concerning uniforms. From then on, the wearing of "antique" leather jackets and helmets was prohibited, and the new nylon jackets appeared. The L-1 was the first of these. It was generally cut like the A-2 with collar, cuffs, and waistbands in wool knit. Olive green nylon became the new color standard. The L-1 was quickly replaced by the L-2, with the same shape, but the color standard was changed to dark blue. A few months later, nylon flight suits would appear. Then the L-2 was transformed into the gray-green nylon L-2B with a green lining. Color standards had changed again to sage green. The Avirex USAF L-2B is still sold (through the Cockpit for example) and the fashion experts have nicknamed it "the nylon bomber". While still a classic, it could never rival with the old horsehide A-2 flight jacket in prestige.

Even though the new Air Force had its new jackets, leather had not had its last word. In June, 1950, the Korean war would bring to a head the conflict between leather and nylon. At the time, the U.S. Air Force fighter squadrons, made up mostly of Lockheed F-80 Shooting Star jet fighters, were based in Hawaii and Japan. The U.S. was forced to defend with World War II vintage P-51 Mustangs based in Korea. The Mustang pilots were almost all experienced fighter pilots from the World War II Pacific and European theaters. In those first tough months, they fought in the old way, equipped with their A-2 flight jackets in leather, and even with their leather helmets and goggles. Very quickly, Air Force staff realized that

Brigadier General Charles "Chuck" Yeager, the fabulous test pilot with an almost interplanetary reputation, at Edwards Air Force Base in California, photographed in front of an F-15, the hottest fighter in the Air Force inventory.

the P-51 Mustangs could not rival the Soviet Mig jets, and brought aircraft carriers into the war. Thus it came about that the "Valley Forge" headed for Korea with the first Navy jets: F-9F Panthers as well as World War II vintage propeller-driven Sky Raiders and Corsairs aboard. The U.S.A. was fighting a defensive war alone until fast jets could be brought into the conflict. Aboard the aircraft carriers leather was still the fashion. Esprit de corps was high even in the tough winter of 1950. The Navy pilots wore their old G-1s with crash helmets painted in Navy gold. Nylon just couldn't hack it! But, as of 1951, General Douglas MacArthur, overall Commander of the U.S. Pacific Forces, was able to convince Washington to commence an offensive war in Korea, beginning with the brilliant landing of American forces at Inchon, behind enemy lines. In support of this offensive, newer F-84 ground-attack jets came into battle, followed soon after by the extraordinary North American F-86 Sabre jets. It was over Korea that the first jet air battles took place, as American Sabres blasted Russian Migs out of the skies at a kill ratio of 12-to-1.

Jet air combat was here to stay and for once the rules and regulations concerning uniforms were strictly enforced. No one was allowed to wear a leather jacket or helmet. Standardization and survival were the keywords for 1951! The jets were hermetically sealed, pressurized, equipped with air conditioning, and flew at the speed of sound. The Air Force pilots had to give in and wear their nylon flight jackets, and flight and anti-G suits. Meanwhile, the Navy pilots still proudly clung to their leather G-1 jackets. But for the very high altitude flying and combat taking place 30,000 to 50,000 feet above sea level, no one needed a leather jacket...

"Life" Magazine, more than any other publication, covered the Vietnam war in its entirety. On February 25, 1966, it published this cover showing jets taking off on a mission from Da Nang.

Only a few privileged pilots such as the test pilots at Edwards Air Force Base in California were still permitted to wear leather. Chuck Yeager was part of that group. Nylon had won its first battle! Yet in memory of the good old times the Air Force pilots could still wear the special white or checkered silk scarf. That was the only concession permitted to nostalgia. Thus, from the middle of the Korean war onwards, leather was banned from the Air Force, and no pilots were permitted to wear leather flight jackets until August 15, 1987. On that date, the U.S. Air Force reintroduced the leather A-2 flying jacket for official active duty wear, and Avirex was asked to help in perfecting the new specs and ultimately supplying the first few hundred leather A-2 jackets to be issued to the best Air Force pilots, those who had "the right stuff". An important day, since it coincided with the 40th anniversary of the Air Force...

Between 1953 and August 15, 1987, the war of the uniforms would rage on. The objective was to go twice or three times as fast as the speed of sound and to break Mach 2 or Mach 3 and beyond. From that time on the pilots became a protected species whose life was even more precious than their multi-million-dollar planes. Years were needed to train a good pilot, not even taking into account the cost, also many millions of dollars. Uniforms and flight suits became more and more sophisticated, with special survival vests and equipment. The pilot would need to be able to save his skin in all hostile environments: jungles, swamps, seas, deserts...

In 1964, at the start of the Vietnam conflict, appeared the first flight jackets and flight suits made of nonflammable nylon, which were only standardized in 1967. These were the CWU 27 Type, which the pilots disliked because they were very hot to wear. At the time a joke circulated around the bases concerning these flight suits. The pilots agreed that in case of fire they would die from asphyxiation rather than from other causes... They were very happy about that because their bodies would remain untouched, protected by these marvelous nonflammable nylon flight suits!

Yet certain special groups took liberties with the uniform discipline. During the Vietnam war special operations groups were secretly based in Northern Thailand and Laos to fight the Communist North Vietnamese. They were the First Air Commando Group, flying propeller airplanes (Types 0-1, 0-2, B-26K, A1-H and T-28D), and another group nicknamed "The Ravens", as well as the CIA's Air America, also flying a variety of prop-driven aircraft. The majority of these pilots had locally hand-tailored camouflage flight suits in cotton, with a tiger stripe pattern. Meanwhile, Navy pilots continued to wear their G-1 leather jackets although flight operations required the use of fire-proof nylon garments.

Unfortunately, after Vietnam came the period of disenchantment. Many Air Force pilots quit the service totally disgusted. Many entered private life and became airline pilots. They were missing an esprit de corps, partially caused by the war and the political restrictions placed on air operations by civilian leaders that wasted American lives and prolonged the war, as well as just not flying enough in the postwar environment. The era of bureaucracy had arrived. Spiritually, the pilots missed their "leathers".

"It would almost seem that we had to wait for the "Top Gun" film for the Air Force to contact Avirex and renew its interest in leather", comments Jeff Clyman, President of Avirex. "Can you imagine the incredible power of leather? Today, pilots wearing animal skins are in command of ultra-sophisticated airplanes such as the B-1 bomber, each worth hundreds of millions of dollars and able to destroy whole cities. It is not a matter of misplaced machismo, but of some vital need to wear the genuine old leather flight jacket. The A-2 invented in the thirties represents the real spirit of aviation. In 1988 you can't just talk of "the right stuff": you must also mention the pilot's second skin, the leather flight jacket. ■

Charles "Chuck" Yeager, cowboy of the sky, a truly patriotic American whose definition of courage is simply doing your duty...

Opposite: Chuck Yeager as an 8th Air Force fighter pilot in Europe, in front of his P-51 Mustang with his ground crew. An ace with more than ten kills, he and his Mustang, "Glamourous Glennis", succeeded in shooting down one of the first Nazi jets, the Messerschmitt ME-262. Centre left: One of Yeager's favorite photographs, wearing his cowboy hat — a picture of a free nature lover. At left: A portrait of Yeager during World War II: Tranquil and with a twinkle in his eyes: evidence of character and "the right stuff". Above: A picture of the Bell X-1 Rocket Plane which he would also call "Glamourous Glennis" and in which he became the first man to break the sound barrier on October 27, 1947. Inset: The leather A-2 "Right Stuff" jacket manufactured in homage to Chuck Yeager by Avirex, and now a collector's item. Note the embroidered X-1 Rocket Plane on the pocket flap.

Far right: Chuck Yeager and Jeff Clyman, President of Avirex, at Edwards Air Force Base in Palmdale, California, in front of a Northrop F-20 Tiger Shark. Clyman is also a jet pilot... Two center pictures: When film meets reality. These two pictures were taken at the legendary Californian bar and pilots' "watering hole", Pancho Barnes's cafe, a hangout for Yeager and his colleagues during the forties and fifties. Above: The real Yeager in front of the cafe. Below: Still at Pancho's but with Sam Sheppard, who played the role of Chuck Yeager in the unforgettable movie by Philip Kaufman, "The Right Stuff", taken from the remarkable book by Tom Wolfe. Above: Yeager again at Edwards after a flight aboard an F-20. Top far right: A detail of the Right Stuff leather jacket by Avirex with the X-1 logo.

Pancho Barnes's cafe, the Californian desert "watering hole", home for Yeager and his colleagues. . .

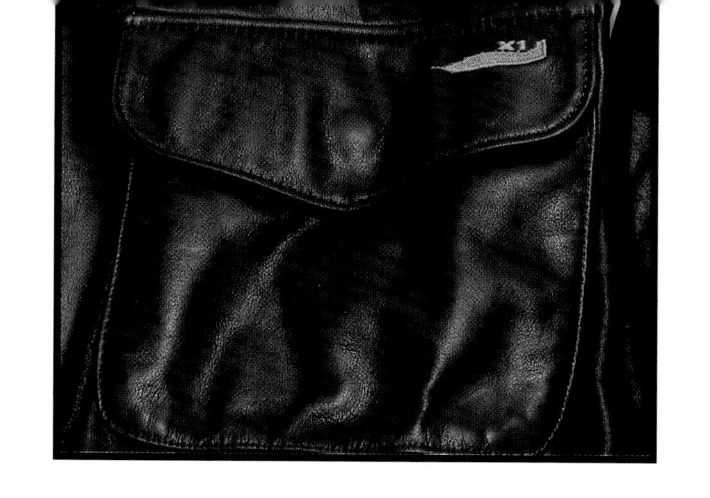

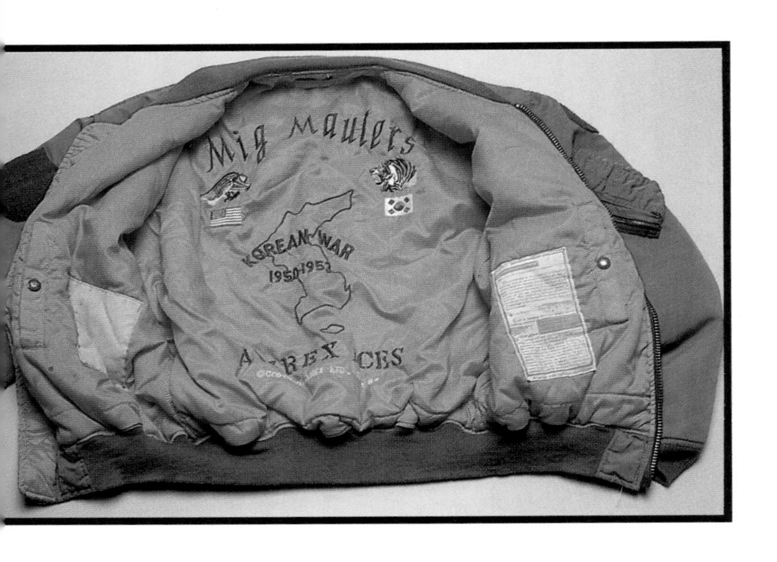

An embroidered jacket in honor of Korean veterans.

Far left: The hand-embroidered back of a classic Avirex athletic jacket (leather and wool). The embroideries represent a soldier's map of Korea surmounted by a freestyle tiger. This line was inspired and created in memory of the American veterans of the all-but-forgotten Korean war. Upper left: A picture taken from the movie "The Hunters", filmed in 1958, in which Robert Mitchum was the star Airforce ace fighting Communist aggression in the last aerial battles in the Korean skies. Lower left: The pilot in the middle is wearing a B-15 with fur collar. The other two are wearing L-2s with knit collars. Above: A Sabre jet on a Korean air base runway during maintenance. Certain aviators wear parkas nicknamed Admiral Byrd parkas in honor of Admiral Richard Byrd. The others are wearing USAF Type N-3B parkas, and World War II style wool-lined "Mackinaw" coats. Top far left: The lining of a B-15 flight jacket made by Avirex, also showing a pilot's map of Korea.

*When I die I'll go
to heaven 'cause
I've spent my time in hell...*

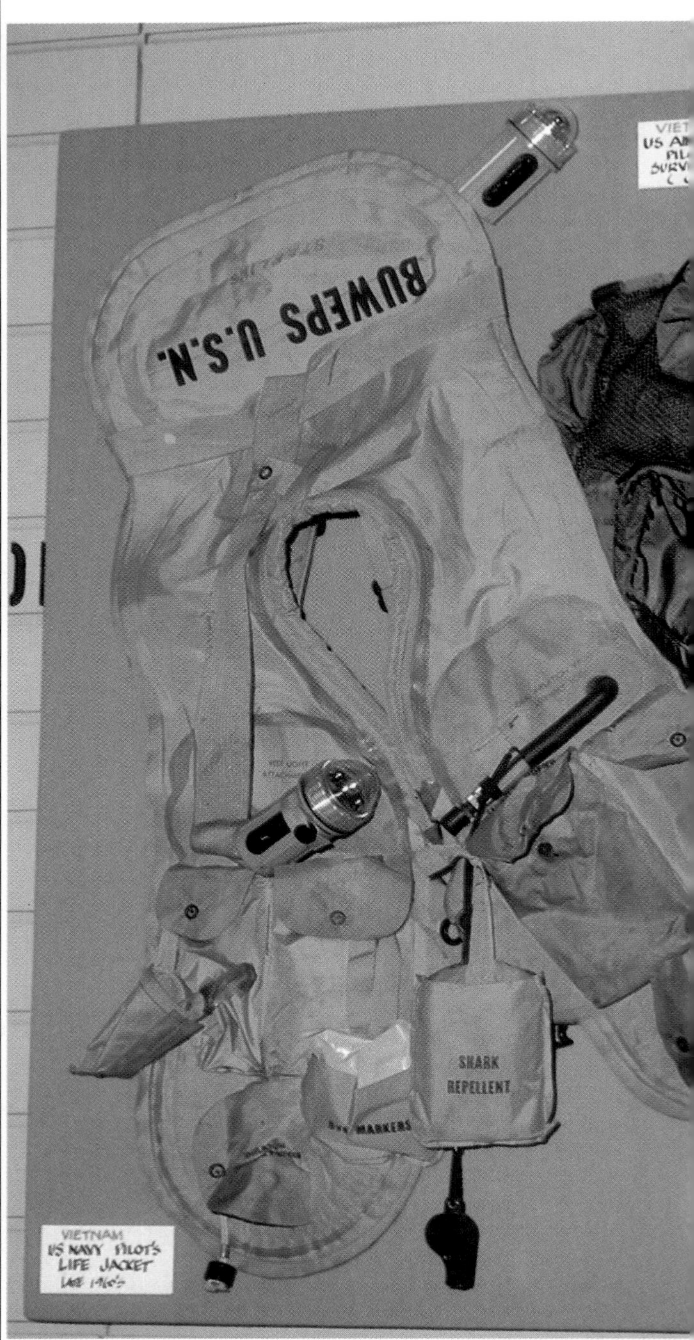

***T**op Right: An Avirex jacket in black velvet produced in very limited quantities in honor of Vietnam veterans. The back and sleeves are embroidered by hand. The map of Vietnam and dragons are capped with the wards "First Air Cavalry". This unit was that of Frank Marchese, Jeff and Jacky Clyman's partner. Centrefold: The famous "R & R" jacket, and a real artifact from the Vietnam war, hand-embroidered with the legendary saying: "When I die I'll go to heaven 'cause I've spent my time in hell", as well as a pilot's Mae West life jacket (1964) and different items used by American pilots in "Nam". Bottom left: On October 20, 1967, "Life" published a report on American prisoners of war in Vietnam. Let us always remember that there are still MIAs and POWs over there, no matter how hard certain government circles want to shut the door on that page of U.S. history... Top left: A Republic F-105 "Thud" photographed with infrared film in 1967 on an American base in South East Asia.*

the leather flight jacket would be sold by the carload worldwide.

***A** series of photographs by George Hall, one of the best U.S. aviation photographers... Far left: Sunset in California on an F-14. Lower left: Tom Cruise in the movie "Top Gun" standing up in the cockpit of his F-14 (also a Cockpit catalog cover) and a U.S. Navy "Plane Captain" telling the pilot to retract the flaps on a jet. The impact of the "Top Gun" movie was such that the American Air Force started as of 1987 to reissue leather A-2 flight jackets, using the Avirex specification. Avirex had, as a matter of fact, also worked with Paramount for the "Top Gun" film before and after the movie came out, and developed another version of its already successful commercialized version of the Naval aviator's flight jacket. Top far left: Top Gun label for official licensed Avirex Top Gun jackets made in 1986. Above: The commercialized version of the Naval aviator's leather jacket manufactured by Avirex, with its original-style patches. This jacket had such an impact all over the world that it is considered an international Avirex classic design. A real hero's leather jacket!*

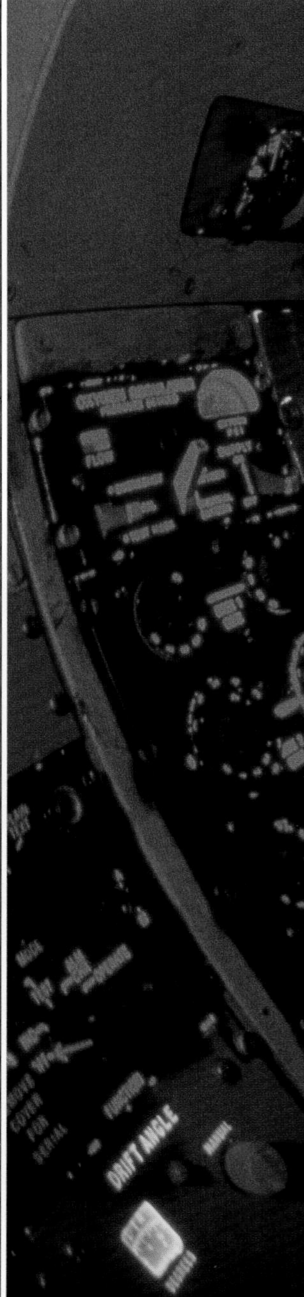

Far right: A picture taken in the inner sanctum, the strategic center of the authentic Top Gun Fighter Pilots' School. Upper right: F-14 Tomcat pilots returning from a mission. Lower right: The Navy Type G-1 flight jacket belonging to Jeff Clyman covered with patches, including an Arizona patch (Clyman was born in Tucson, Arizona). Above: An F-14 during a training flight in the Pacific. The pilot is controlling his turn with extreme precision so as to be heading straight for the deck of the aircraft carrier. Top far right: The new logo for the Avirex Cockpit store on Broadway, international meeting place for fanatics of flight jackets and the world of aviation. Welcome to the club!

"The Right Stuff" and "Top Gun" were the hit movies of the eighties' Jet Age...

W A R N O S E

War noses are decorated airplane noses, the best examples of which were done during World War II and the Korean War. If the paintings on the back of aviators' jackets were nicknamed "war hand-paintings" those on the noses of war bird aircraft were defined as "nose art". In this typically American military art, the sources of inspiration were infinite. Naturally, pinups were in the lead, followed closely by cartoon heroes, and girl friends. It was fun art which was used to make them forget momentarily the seriousness and tragedy of war. Very often these nose art decorations were considered as good luck charms, able to trick death and cheat destiny. Far left: The pose and Manuki's telephone were pretty explicit. This young lady was painted on a B-29 in Korea just as "Our Gal" below, which belongs to the 19th Bomb Group. Above: The grim reaper with his sickle ornament the nose of this Mustang P-51 and helmet of its pilot, and the airplane is appropriately nicknamed "Death Rider". The nose and motors of this Flying Fortress B-17, based at Chino Airport, California, and shown at left, are in themselves the most beautiful of war noses. As for Clark Gable, he was in reality a gunner on a B-17 in the 8th Airforce. Here he poses next to the B-17 "Delta Rebel" N° 2.

*"**B**etty Grable", who is on the nose of a B-25 at Chino airport, was as well endowed as all the pinups on this page. More women! Always women! "Lucky Strike" — not the cigarette! — was painted on the nose of a B-24. The beauty on her cloud rested on the front of a small L-5 reconnaissance plane. The two beauties "South Sea Sinner" and "That's It" adorned the noses of two B-29s of the 28th Squadron based in Korea. Nose art disappeared soon after the Korean war, to reappear during Vietnam, and then disappear once more until 1988, when the U.S. Air Force permitted, along with A-2 jackets, the readoption of nose art. Today, the most beautiful collection of nose art is to be found in museums such as that at Chino Airport in California, The U.S. Airforce Museum in Dayton, Ohio, The National Air and Space Museum in Washington, D.C., and The Confederate Airforce Museum in Harlingen, Texas.*

W A R   N O S E

"Passion Wagon", photographed during the Sun and Fun air show in Lakeland, Florida. It is during such air shows as this, or as the Valiant Air Command one of Tico in Titusville, Florida, that one can admire the most beautifully decorated airplanes. But the best can still be seen in Oshkosh, Wisconsin, where hundreds of airplanes and thousands of airplane buffs meet each year at the end of July. "Atomic Tom" and "Mule Train" were two famous B-29s based in Korea. "You've had it" was the nose art of a B-17 of the 8th Airforce, whereas the Death Mask wearing a Special Forces beret belongs to the nose art of the Vietnam war. This airplane belongs to a former member of the 5th Special Forces Group, Hanoi Recon.

WAR NOSE

W A R   N O S E

HAND HOLD
U. S. ARMY
AIR CORPS SERIAL NO. 49-3829
HYD. FLUID TANK
GROUND
Double trouble
14

"Double Trouble" is the name of the T-6 belonging to Jeffrey Clyman, President of Avirex Ltd. A pilot whose life is devoted to aviation, Clyman sponsors, through his Cockpit catalog and store "The Cockpit", a precision flight team of T-6s piloted by Dan Dameo, Dan Calderale, Bill Dodds, Jerry Walbrun and Chuck Kruger. The metallic gray paint on "Double Trouble" is original (the airplane has an actual combat history). The witch on her broomstick was originally painted on a P-51 Mustang in Europe in World War II. At left: Chuck Kruger and his "Chuck Wagon" T-6. The "Mission Inn" B-29 was based in Korea and belonged first to the 22nd BG and afterwards to the 19th BG. The "Our Gal" pinup appears on three different airplanes: a B-24 and a B-17 in World War II, and a B-29 during the Korean war. Fortunate woman! The gunner wearing his B-3 is in the middle of painting the 100th bomb, symbolizing the 100th mission on a B-17 of the 379th BG.

AT-6D-NT
IAL NO. 49-3829
400 LBS.
CLYMAN J M

AVIREX

In ten years, Avirex has sold more than 2,500,000 jackets all over the world. The totality of these jackets, in leather, nylon and cotton, if stretched arm-to-arm, would create a gigantic air bridge between New York and Paris. More symbolically, this incredible chain of jackets linking America to Europe would retrace the route taken by Charles Lindbergh from Long Island to Le Bourget. From the U.S. to France, via Japan, Korea, Hong-Kong, Italy, Germany, and Austria, not to mention Denmark, Belgium, Spain and England, the Avirex "planet" gives work to 16,270 people. At this level, it can no longer be called a success or an explosion, but rather a religion or even a philosophy. It is a wager made and won by Jeff Clyman who, since 1975, has set his sights very high by adhering to a master plan: to manufacture the jackets worn by World War II pilots. But, above and beyond just plain manufacturing, the objective of the endeavor was to keep alive the very special spirit of the pioneers and heroes of U.S. aviation, such as Lindbergh, Doolittle, Fisher, and Yeager. Mission accomplished! It was also Avirex which supplied the Mil Spec jackets for the movie "Top Gun", and which had already produced a "Right Stuff" line of clothes for the launching of the movie of the same name. For many years the brand has been on garments manufactured for the current American Air Forces, and the ambition of being the first company in 45 years to supply the U.S. Air Force with A-2 leather flight jackets has been achieved. The fabulous Avirex spirit is also found in their New York store, the Cockpit, partly conceived as the deck of an aircraft carrier, and in which is found a real North American Texan AT-6 World War II aircraft! For millions of its fans, Avirex really has "the leather and the 'right stuff' which makes us all heroes". A perfect example of a modern American dream come true!

IN SEARCH OF THE PAST

Opposite: The symbol of Avirex: Jeff Clyman's antique T-6 flying over the skyline of Manhattan. The meeting of the past, the present, and the future. Insets: Gag picture in Clyman's museum-office: Jeff, Jacky, his wife, and his partner Frank Marchese up to the neck in their own products. The President of Avirex in the cockpit of this T-6 before takeoff.

“For years I spent my time on the track of the old leather jackets worn by American pilots in World War II. After interminable research I unearthed, one by one, all the secrets of how to manufacture the authentic cult bomber jacket of the American hero. But the most difficult part awaited me: to share my passion with millions of others... and make a living doing it!”
Jeff Clyman was born on December 24, 1945, in Tucson, Arizona. His father, Major Martin Clyman, who had been a flyer since 1932, had been a pilot, in the 20th Air Force Bomber Command, of the B-29 “Nightmare”, and had become a surgeon and research physician after the war. During his childhood, Jeff was enthralled by stories about General Chennault's Flying Tigers, who had done battle in China against Japan. His father, who had been based in Chungking, Cheng Tu and Kunming, recounted numerous exploits of these legendary flyers as well as the legends of the pioneers responsible for the first air bridge in history, between India and China, the famous “Hump”. Not forgetting the raid from Rangoon to Bangkok and all the adventurers of the sky who braved death aboard their Curtiss P-40s with their shark teeth, wearing their magnificent A-2 or B-3 jackets. In addition, he was told other stories by his uncle, Eugene Grant, who had been a fighter pilot on the European front.
Even as a toddler, he dreamed of flying and becoming a pilot. At ten he was already riding in the old Stearman cropduster based on the ranch next to his grandfather's farm. As a teenager, his dream was still strong. He entered the University of Vermont where, in addition to his studies, he practiced speed skiing and excelled in AF ROTC. In 1967, he got his bachelor's degree, his pilot's license, and a skiing medal. Right after that he was posted to Central America.
His problem was that he wanted to be a pilot but not spend his life in the armed forces. One important detail: since his teens he had worn the original horsehide A-2 flight jacket which his father had given him.
By 1971, he found himself back in Washington, D.C., where he managed to obtain a law degree and met his wife, Jacky, a French-American who was stationed there as a State Department interpreter. To be able to fly, Jeff regularly attended all the air shows in the U.S. After their marriage, Jeff and Jacky settled in New York where Jeff worked for a prestigious law firm.
Both of them were nicely established: Jacky worked in the international department of a major bank and two little Clymans were soon born, Scott and Shawn. Despite this, Jeff was bored to death. He still flew, but less and less frequently and not in any military aircraft. After about two years he left the law firm and went to Columbia University, where he earned his master's degree in business administration. But he continued to yearn for something more. He was a product of the wide open spaces: roaming the deserts on horseback, the mountains on skis, and the skies in airplanes... but he still needed the thrill and the challenge of adventure. He did not know then that he had been born under a lucky star, which made the legendary “American dream” come true for a handful of privileged people.
As in all success stories, Jeff's would come somewhat by chance. Every weekend he would hit the airports and aviation hangouts wearing his famous bomber jacket. This museum piece always intrigued his friends, who would ask him where he had bought it. After the 500th time, Jeff decided that he would manufacture the jackets worn by the American pilots during World War II and which had not been issued since 1945. Not just an imitation, but the real McCoy down to the last stitch. He did not have the means, and knew nothing about his new endeavor. Far from discouraging him, his wife, Jacky, helped him; her parents, both French, had been associated

Jacky and Jeff in Titusville, Florida, during the Valiant Air Command show in front of a P-40 Warhawk. Their slogan, "authenticity", is also that of their store and catalog, the Cockpit.

with the American Air Force since the war and therefore Jacky had grown up on U.S. bases in Morocco, Spain and America. A real family of cowboys of the sky! Both Clymans were to learn as they went along. He himself tells of his apprenticeship, his ambitions, his failures, his successes, and his joy at finally being able to do something which thrilled him.
“In 1975, I borrowed the money to start Avirex. Then I went to work. I poured over countless phone books to find the old army suppliers. Many had disappeared. I needed patterns, catalogs, advice. Where could I buy zippers, leather, sheepskin, thongs? The further I progressed, the more problems I found! The pilots in World War II had been smaller and slimmer than the present generation and my first jackets were too tight. They did not fit. We had to start over and over again. The more I learned, the more I realized I didn't know. What was the exact finish on the B-3 jackets? I had the impression that the manufacturing secrets of these legendary flight jackets had been lost.”
Jeff would go all over the country looking for tanneries and leather suppliers. He wanted to make

authentic jackets. It is his sole philosophy. In 1979, during a fashion show in New York, he met Frank Marchese, who was later to become his partner. Frank was a Vietnam veteran who had been manufacturing, near San Francisco, marvelous Western-type leather jackets with fringes, to sell to boutiques such as Northbeach Leather and Pithic. The two leather fanatics worked together but at the time the U.S. market did not understand their products. In contrast, Europe, especially France and Italy, and Japan, were already crazy about Avirex products, which they had discovered during a trade show in New York. Jacky was also now working with Jeff, as his partner.

At the beginning of 1980, the U.S.A. was finally

© Mark O. Greenberg

The Avirex staff in Long Island City brought together for a quick picture. They all participate in the Avirex "trip": "Jackets made to last generations".

beginning to recover from the crisis of the post-Vietnam war years. Part of this recovery involved a return to traditional values and attitudes reminiscent of the forties'. Patriotism was no longer out of fashion, and people took refuge in historic authentic values represented by images of America's pioneers: Western proneers — sky pioneers. The feeling of strength was just beginning to return after the disastrous Carter administration had been succeeded by the Reagan one. America was on its way up again. At that time, Jeff finally found a tannery to work with him in perfecting the process to "antique" sheepskin, to give it that authentic look. A mistake was made in the formula during production and Jeff thought the game had been lost. But no! On the contrary, he had just reinvented the famous old and worn looking leather! He had won the battle.

At least almost!

Hollywood later adapted Tom Wolfe's book, "The Right Stuff". Avirex was to manufacture the jackets while Sam Sheppard starred on screen as the American hero Chuck Yeager, the man who broke the sound barrier, on October 27, 1947, aboard the X-1 prototype. Jeff Clyman and Chuck Yeager became friends, and Avirex produced a "Right Stuff" collection for an AC Delco promotion with Chuck Yeager as the spokesman. Avirex had gained some lift under its wings! At about this time, Avirex undertook a number of important Department of Defense contracts to manufacture current-issue nylon and Nomex jackets.

The same phenomenon was to reproduce itself a few years later with "Top Gun", the modern aviation film epic. Avirex was again involved. This time the impact was even more powerful and their aviator-style leather jackets streamed around the world. All the while Jeff, Frank and Jacky hoped that one day the Air Force would bring back the leather flight jacket, symbol of the World War II flying heroes, and that Avirex would be directly involved. This dream became reality, for in 1987 the U.S. Air Force reissued the A-2 leather pilot jacket, with Avirex helping to design the new specs and winning the first official contract since 1943. Avirex worked with the Airforce Logistics Command at Wright Patterson Airforce Base as well as with the Tactical Air Command in developing an acceptable specification which would continue the historical tradition of the jacket while permitting large-scale production to suit the needs of the Airforce.

Avirex had finally found and perfected for the modern world the manufacturing secrets of the A-2 and B-3 bomber jackets. Clyman, the diabolical alchemist, the bored boss/lawyer, the flying aficionado, had won his battle!

His first stroke of genius had been to discover and perfect antique leather with Frank Marchese. The second came from Jacky when she convinced him to open the Cockpit retail outlet, combining the Avirex spirit and mission: authenticity in a hard-core aviation spirit. In 1986, the Cockpit moved a few blocs further down on Broadway to become one of the world's most original boutiques! Imagine, if you can, the decks of a giant aircraft carrier with a real AT-6 Texan, painted in bright yellow instrument training colors, at the end of the runway. In the superstructures of the carrier are mannequins wearing all the different bomber jackets, display after display of original, authentic jackets as well as artifacts, all rare and splendid! Jacky, Jeff and Frank had been wise enough to choose as their principle source of inspiration the genuine jackets worn by the famous American heroes: Chennault's Flying Tigers' jacket, Chuck Yeager's A-2, Patton's B-3, Eisenhower's "Ike" jackets, and hand-painted bomber jackets showing movie stars and singers who had been sent on U.S.O. tours to build up troop morale such as John Wayne and Glenn Miller, as well as the war paint designs, the pinups, the flight jackets worn in Korea and Vietnam...

In 1986, seeing how enthusiastic the French were about Avirex, they opened a subsidiary company in Paris with another aviation fanatic, André Leguen. By 1988, Avirex had sold 2,416,000 flight jackets all over the world and supported 16,270 people on this planet. But Clyman, the cowboy of the sky, never forgets that his "mission" was won with the aid of his partners, Jacky and Frank, his staff at Avirex, and his international associates. ■

jeff Clyman and Frank Marchese, two indomitable fanatics. . .

***W**elcome to the Avirex universe! At left: Jeff Clyman and Frank Marchese at Linden Airport in New Jersey in front of "Double Trouble", Clyman's T-6. Both are committed buffs: the first is a pilot of war birds and jets, the second did a tour in Vietnam with the air mobile First Air Cavalry. Bottom right: The spirit of Avirex is captured in this nostalgic collage: A woman pilot, a B-17 crew, an old bottle opener, a World War II match book with an old scribbled phone number... Above right: This G-8 Type, open cockpit leather jacket is one of the basics in the Avirex-Cockpit collections. Above: A giant billboard for the Avirex brand which dominated La Cienaga Boulevard in Beverly Hills, Los Angeles.*

The Avirex products are meticulous replicas of the legendary A-2 flying jackets. . .

A*t left: Jeff Clyman and Frank Marchese in the full swing of creation in Jeff's "boss's lair". Each authentic jacket has been dissected and recorded. Today, the Avirex products are authentic reproductions of the legendary flying jackets. Everything has been respected, down to the last stitch. No detail has been forgotten in this incredible work of investigation and study: a universe and style respected to the letter! Above: Joseph d'Anna in the Avirex showroom in front of the new collection, and the superb Type A-2 flying jacket which brought success to the brand. It was conceived to last for ever. Bottom right: Jeff Clyman doing pre-flight inspection of his T-6, "Double Trouble".*

the Defense Department has worked with Avirex for many years...

It is not by chance that Avirex has become the N° 1 in the world for leather jackets, and it is not by chance that the American Defense Department has worked with this company for many years. Their tanneries and factories work in the traditional way, attentive to the most minute details, from the linings to the cuffs and waistbands, as well as the zipper and the thongs. The choice of sheepskin leather is just as rigorous. For months, Jeff and Frank tracked down the old suppliers, finding the old catalogs, the original patterns, piercing the secret of these bomber jackets "built" to last for generations.

At left: Two great garments bearing the Avirex trade mark: the only thing missing is the actual patina of wear — "Keep 'em flying!" Other pictures: One of the numerous Avirex factories, with Jeff Clyman at work in the cutting room. Nothing escapes the boss's eye and his employees have created a sign in his honor: "Achtung! I am watching you!" Whatever happens, humor occupies an important position in the company...

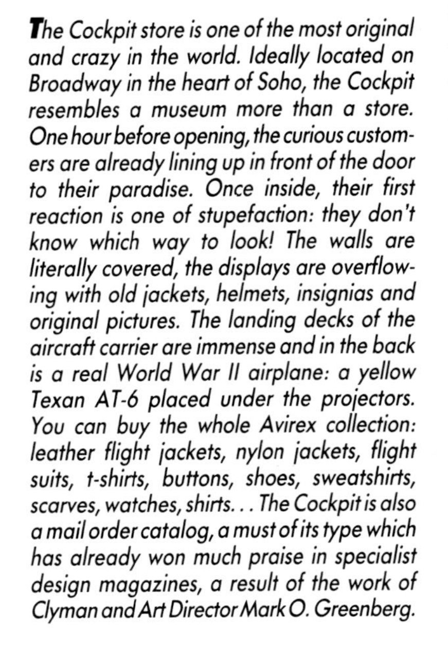

The Cockpit store is one of the most original and crazy in the world. Ideally located on Broadway in the heart of Soho, the Cockpit resembles a museum more than a store. One hour before opening, the curious customers are already lining up in front of the door to their paradise. Once inside, their first reaction is one of stupefaction: they don't know which way to look! The walls are literally covered, the displays are overflowing with old jackets, helmets, insignias and original pictures. The landing decks of the aircraft carrier are immense and in the back is a real World War II airplane: a yellow Texan AT-6 placed under the projectors. You can buy the whole Avirex collection: leather flight jackets, nylon jackets, flight suits, t-shirts, buttons, shoes, sweatshirts, scarves, watches, shirts... The Cockpit is also a mail order catalog, a must of its type which has already won much praise in specialist design magazines, a result of the work of Clyman and Art Director Mark O. Greenberg.

The Cockpit store is one of the most original and crazy in the world...

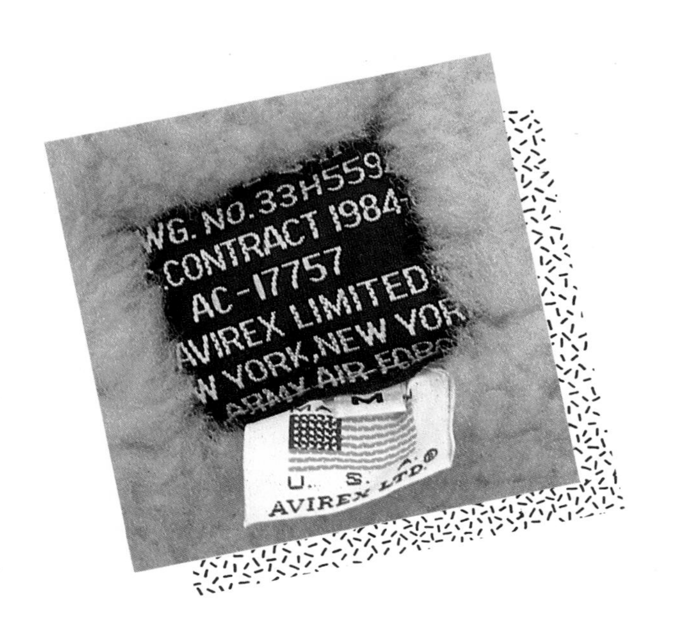

the images and the world of Avirex are there: the past, present and future. .

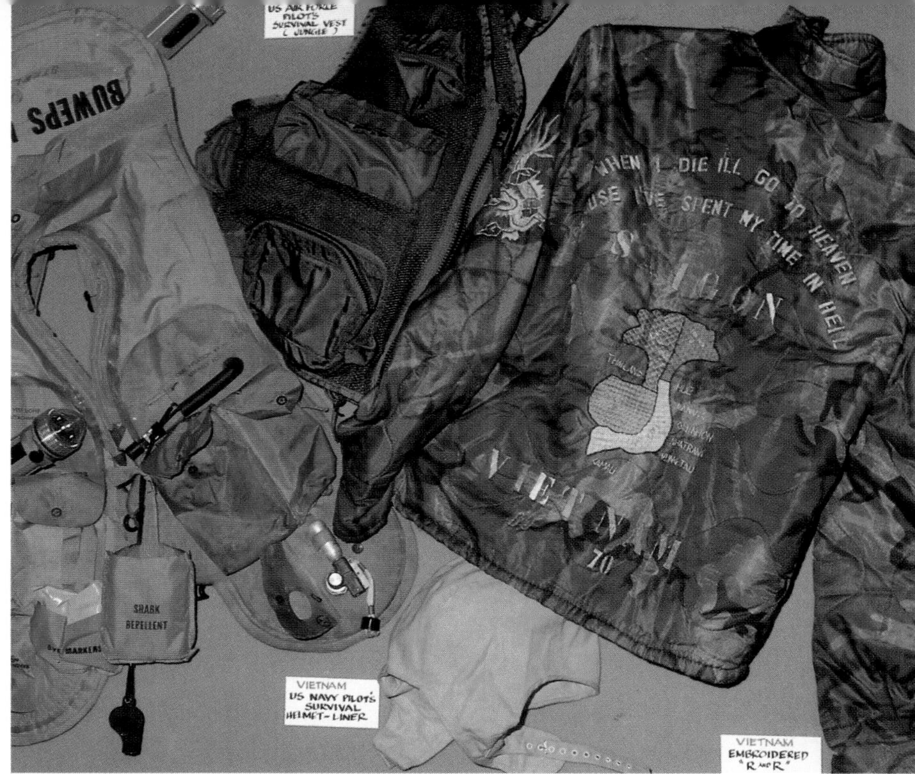

Let us penetrate the inner sanctum of the Cockpit to better understand the universe and the goals of the dedicated aviation buffs who comprise Avirex. The Cockpit is the most explicit "window" representing Jeff and Jacky Clyman's slogan, "Authenticity". The images and the world of Avirex are present in their entirety: the past, present, and future. There, in a show case, are a German cloth aviation jacket and an American flight jacket. Each display resembles a surrealistic canvas, painted with love. There, a slogan embroidered on the back of a jacket worn by a U.S. soldier in Vietnam saying "When I die, I'll go to heaven, 'cause I've spent my time in hell". Opposite, a test pilot of the Jet Age welcomes the visitors, immobile, as if paralyzed by the force of the jets. Raise your eyes and the pilot-mannequins in flight suits are watching you, up there in the superstructure of the aircraft carrier. Welcome to the aviation fanatics' club!

We worked for two days to create this first real flying jacket image. . .

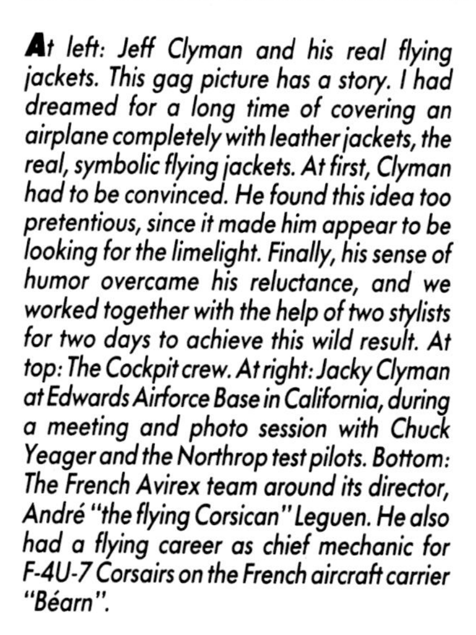

A*t left: Jeff Clyman and his real flying jackets. This gag picture has a story. I had dreamed for a long time of covering an airplane completely with leather jackets, the real, symbolic flying jackets. At first, Clyman had to be convinced. He found this idea too pretentious, since it made him appear to be looking for the limelight. Finally, his sense of humor overcame his reluctance, and we worked together with the help of two stylists for two days to achieve this wild result. At top: The Cockpit crew. At right: Jacky Clyman at Edwards Airforce Base in California, during a meeting and photo session with Chuck Yeager and the Northrop test pilots. Bottom: The French Avirex team around its director, André "the flying Corsican" Leguen. He also had a flying career as chief mechanic for F-4U-7 Corsairs on the French aircraft carrier "Béarn".*

avirex and the Cockpit have become the Bible of the world of aviation and flight jacket fanatics. .

In a few years, Avirex and the Cockpit have become the Bible of the world of aviation and flight jacket fanatics. But Clyman knew this "mission" was successful when the veterans and survivors of past wars started to write to him and compliment him on his products, and more especially on his "spirit", and sending him original photographs. This correspondence is so important that it is, today, included as a special chapter in the Cockpit catalog. At left: An evening with Mark O. Greenberg. We put together this still life that we naturally baptised "The Cockpit Spirit". At right: Living legends! A jacket in sheepskin from the latest Avirex collection inspired by English bomber pilots. Above: The cover of "The Saturday Evening Post" of May 31, 1941, seven months before Pearl Harbor, already had as a title "Showdown in the Pacific". A harbinger of days and years to come!

AVIREX PLANET

In ten years, Avirex has sold 2,416,000 leather and textile jackets throughout the world. If these jackets were laid side by side, they would create a gigantic air bridge between America and France. This number is doubly symbolic since it corresponds to the route flown by Charles Lindbergh. The Avirex planet comprises: the U.S., France, Japan, Korea, Hong-Kong, Italy, West Germany, Belgium, England, Austria, Denmark, and Holland. It is a wonderful adventure which is also a journey back in time to the authentic World War II aviation universe. Today, the Avirex planet supports directly or indirectly 16,270 people and thrills thousands of "hard core" leather buffs. An American leather dream come true!

This map appeared on Monday, April 6, 1942. It pinpoints the news along the international fronts during the week from the 1st to the 6th of April, as well as reporting on the natural resources of the forces present there. The Avirex logos superimposed on it indicate manufacturing and distribution points of Avirex products.

1 PHILIPPINE TEMPO RISES

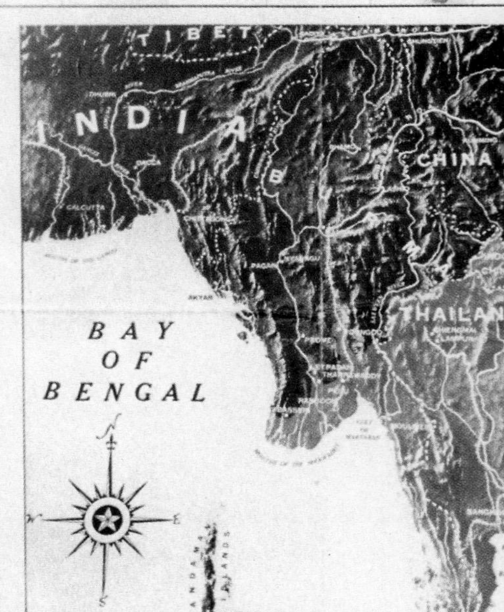

2 BURMA, THE BULWARK AGAINST JAP STABS AT CHINA, INDIA

BATTLES DEADLOCK

...ate effort at compromise seen

8 DESERT FORCE ORGANIZED

Will train under combat conditions

Maj. Gen. Patton

...ACIFIC WAR COUNCIL

...le political, economic aspect

9 ARMY TO PUBLISH PAPER

"Yank" will be run by soldiers

... FERRY DOWN UNDER

...sports not one attacked

10 RADIO SERIES FOR ARMY

Programs will be short-waved overseas

William Shirer

...ER NOTES REDUCED

... censor proves he's human

...HIT SHOW REVIEWED

...p Yaphank" for AER fund

11 PARADES MARK ARMY DAY

Open house for public at air fields

12 MARINES GET NEW PACK

Parts are quickly removable

Marine Battle Uniform

13 AXIS SUBS SUNK TOTAL 28

Additional sinkings not yet announced

14 REPLACEMENTS IN ICELAND

Army relieves Marines there since July

15 BRITISH RAID ST. NAZAIRE

RAF continues long-range sweeps

16 NORSE SHIPS FLEE SWEDEN

Run blockade in dash for England

17 THE NEAR EAST WAITS

American supply missions bolstered

18 ACTION IN THE ARCTIC

Increased convoys point the way

THE VICE PRESIDENT
WASHINGTON

May 2, 1988

Mr. Jeff Clyman
President
Avirex Ltd. - Avirex U.S.A.
33-00 47th Avenue
Long Island City, New York 11101

Dear Jeff:

Thank you very much for sending me the Caterpillar Association Pin and for your kind words. I appreciate your thoughtfulness.

Warm regards.

Sincerely,

George Bush

At right: Major Martin Clyman (on the left), Jeff Clyman's father, a hero of the war in the Pacific and the Hump who was a B-29 bomber pilot. Above left: "Nightmare", Major Martin Clyman's B-29. The numerous bombs painted on the fuselage represent the number of missions, and the camels the number of flights over the Himalayan Mountains. Below: Jeff Clyman's uncle (on the left), standing in front of his Mustang P-51 in France on the European front, shortly after D-Day. Above right: Vice President Bush thanks Jeff Clyman for sending him a special pin commemorating pilots who had to bail out of their planes. Vice President Bush was a U.S. Naval aviator during World War II, flying a Grumman TBF Avenger dive bomber, and was shot down on a bombing raid.

FRANCE

Island
3, rue Montmartre
75001 Paris
Tel. 47.03.37.20

Le Jour d'après
3, rue Poquelin
75001 Paris
Tel. 42.33.66.14

Brook's Company
54, rue de Passy
75016 Paris
Tel. 45.27.19.56

Western House
13, avenue de la Grande Armée
75016 Paris
Tel. 45.00.06.05

Sté Lys
4, rue Pierre-Lescot
75001 Paris
Tel. 42.33.56.30

Diabolo-Menthe
15, rue des Pilliers
Forum des Halles
75001 Paris
Tel. 42.97.57.32

Hémisphères
1, bd Emile-Augier
75116 Paris
Tel. 45.20.13.75

Versioni
84, av. des Champs-Elysées
75008 Paris
Tel. 45.63.86.38

Surplus Neuilly
20, rue de Chartres
Neuilly-sur-Seine
Tel. 46.24.39.46

Guépard
39, av. Paul-Vaillant-Couturier
94250 Gentilly
Tel. 49.86.59.59

PROVINCES

Dakota Line
89, av. Foch
94100 St-Maur
Tel. 43.57.70.90

Surplus américain
18, rue de la Faïencerie
54000 Nancy
Tel. 83.35.49.35

Gold Jean
3, Grande-Rue
26000 Valence
Tel. 75.42.83.72

Jean Bazar
Centre Dauphine
21000 Dijon
Tel. 80.30.63.12

Cripers
4, rue Lanterne
69001 Lyon
Tel. 78.27.14.38

Stock Bellecour
25, place Bellecour
69002 Lyon
Tel. 78.37.17.11

American Graffiti
4, bd Danièle-Casanova
2000 Ajaccio
Tel. 95.50.53.23

India Flip Machine
Place de la Garonne
83990 St-Tropez
Tel. 94.97.18.37

Laine Cannebière
98, La Cannebière
13001 Marseille
Tel. 91.48.43.92

Coconut's Sarl
77, rue St-Jean
62520 Le Touquet
Tél. 21.05.65.30

Aux 1001 Cuirs
13, rue Masséna
06000 Nice
Tel. 93.87.98.73

Surplus Marceau
29, rue Marceau
06000 Nice
Tel. 93.80.51.72

Boutique 103
Passage du Port
83990 St-Tropez
Tel. 94.97.21.26

Western Boutique
5, rue du Président-Faure
74000 Annecy
Tel. 50.45.25.55

Sté Vêtement Loisirs Globe
3, place du Locronan
29000 Quimper
Tel. 98.95.13.74

Globe ST 42424
15, rue du Temple
87000 Limoges

Pénélope
5, rue du Port-de-Castets
64100 Bayonne
Tel. 59.59.12.44

Chapparal
84, rue du Pas-Saint-Georges
33000 Bordeaux
Tel. 56.51.70.35

Sélecto S.D.B.
Centre commercial Colombia
35000 Rennes
Tel. 99.31.39.46

Azur France Distr.
14, avenue Notre-Dame
06000 Nice
Tel. 96.62.23.91

L'Occitanie
3, place de l'Hôtel-de-Ville
11100 Narbonne
Tel. 68.32.16.24

Série Noire
1, rue de la Bourse
59000 Lille
Tel. 20.55.48.77

U.S.A.

Cignal
Century City Center
2828 North Clark St.
Chicago, IL 60657
Tel. (301) 828-1000
(And all U.S. Cignal stores)

Louis of Boston
199 Boylston St.
Boston, MA 02167
Tel. (617) 965-6100

Marshall Fields
11 North State St.
Chicago, IL 60602
(And various stores throughout the U.S.)

Bullocks
925 West 8th St.
Los Angeles, CA 90017
(And various stores)

Saks Fifth Avenue
611 5th Ave.
New York City, NY 10022
(And various stores throughout the U.S.)

I Magnin
Union Square
San Francisco, CA 94108
(And various stores throughout the U.S.)

Chanins
1030 Westwood Blvd.
West Los Angeles, CA 90074
Tel. (213) 208-8686
(And all Chanins stores)

Charivari
441 Columbus Ave.
New York City, NY 10024
Tel. (212) 362-1212
(And all Charivari stores)

Burdines
22 East Flager St.
Miami, FL 33101
(And various Burdines stores)

Macy's Herald Square
Bway & 34th St.
New York City, NY 1001
(And various Macy's stores throughout the U.S.)

JAPAN

Noise Fukushoru-Bunka Co. Ltd.
366 Kamiyamachi, Sijyou-Agaru Nishikiya-Cho, Chukyo-Ku, Kyoto City
Tel. 075-255-1237

Blaze
Sanshita-Dori bldg., 2-7-30 Shita-Dori, Kumamoto City, Fukuoka-Pref.
Tel. 096-352-7876

Marui Young in Shibuya
1-22-6 Jinnan, Shibuya-Ku, Tokyo
Tel. 03-464-0101

Point Forus of Fukudata Clothing Co. Ltd.
Forus 5F 3-11-15 Ichiban-Cho, Sendai-City, Miyagi-Pref.
Tel. 0222-64-8103

Nakata Shoten
6-4-10 Ueno, Taito-Ku, Tokyo.
Tel. 03-831-5154

Empire (Ueno-Shokai)
2-16-10 Sangenjyaya, Setagata-Ku, Tokyo
Tel. 03-422-4500

DENMARK

First Company Store/Vision
Frederiksberggade 24
DK-1459 Kobenhavn K

New York
Kobmagergade 24
DK-1150 Kobenhavn K

Magasin/Depotet
Kongens Nytorv 23
DK-1050 Kobenhavn K

ITALY

Americanino "Made in Europe Stores"
(Throughout Italy)

Nick & Sons srl
Via Cavour
114 Ravenna

Mantovani Sport
Corso A. Pio
65/71 Capri (MO)

Oscar srl
Via Bertani
42 Rimini (FO)

GREAT BRITAIN

Flex
13, Trocadero Center
Piccadilly
London WIV 7 FE
Tel. 01-734-3439

American Classics
400, King's Road
London. SW 10
Tel. 01-352-2833

Street Clothes
38, The Calls
Leeds LSA 7 EW
Tel. 0532-310360

Harrods
Knightsbridge
London SW7
Tel. 01-730-1234

SWITZERLAND

Meyer et Cie
38, rue Monthoux
1201 Genève

Menthalo
16, galerie de Cornavin
1201 Genève

ANDORRA

Pintat Mas
Encamp
Principality of Andorra

BELGIUM

Azul PVBA
Statiestraat 56
02730 Zwinjndrecht

Aquarelle
Century Center
Keizerlei
02000 Antiverpen

Lonneux Wannet
Graaf Van Egnonstraat 7
02800 Mechelen

Samdam
Krommen Elleboog 32
09000 Gent

Salerno NV
63 Abdijstraat
02020 Antiverpen

Three pictures symbolizing the Avirex universe: the past and the present. At the top: An original World War II A-2 flight jacket with a painted back representing a very sexy cowgirl. Jeff Clyman's private collection. At right: A modern cowgirl equipped with a Navy Avirex jacket and a leather knapsack. Directly above: One of the newer examples in the 1988 Winter Avirex collection, the Navy jacket with original patches.

PHOTO CREDITS

Gilles Lhote — George Hall — Fred Sutter — Mark Meyer — Philip Makenna — Mark O. Greenberg —Period photographs: The Charles Lindbergh Museum (Roosevelt Field) — The U.S. Air Forces Museum — The National Air and Space Museum — The Confederate Air Forces.

THANKS

Avirex: Jeff and Jacky Clyman, Frank Marchese, Joseph d'Anna, Mark O. Greenberg, and all the Avirex staff at Long Island City and at the Cockpit store — The Six of Diamonds acrobatic flight team — André Leguen and Avirex France — Carl Scholl and Tony Ritzman from Aero Trader, Chino, California — The Cradle of Aviation Museum / The Charles Lindbergh Museum (Roosevelt Field) — The U.S. Air Force Museum (Dayton) — The National Air and Space Museum (Washington) — The Confederate Air Forces (Harlingen, Texas) — The Chino Airport Air Museum (California) —The Experimental Aircraft Association, Oshkosh, Wisconsin, and the organizers of the Sun and Fun (Lakeland) and the Valiant Air Command (Titusville) air shows — Tex Hill, R. T. Smith, and all the other surviving Flying Tigers — Chuck and Glennis Yeager — Herbert "Fireball" Fisher — Paramount — George Hall — Jim Dietz — Alain Arnhoult Features — The U.S. Air Forces and Navy — Mark Meyer — Philip Makenna — Fred Sutter — Walt Radovich — Chuck Baisden — Chunto — Magazines: "Life", "The Saturday Evening Post" and "Paris-Match" — Les Editions Filipacchi, Paris: Roger Thérond, Patrick Mahé and Nicolas Hugnet — Pierre and Marie Boogaerts — Roger Ritz and Carole Cena — Jean-Pierre Bouyxou — Bill Tappee and Leonore Frye of the United States Travel and Tourism Administration in Paris — Anne Braillard — Jean-Pierre Binchet — Philippe Paringaux —Michel Decron —Pierre Vergnol and Didier Rapot — Michel Sola — Scott and Shawn — Morgane and Donovan — Erika — Patrick de Gmeline, author of the book "Luftwaffe".

THE COCKPIT

The SIX OF DIAMONDS Precision Flight Team Sponsored by The Cockpit Catalog

$2.

For M.C. & VISA Orders in New York Call 1-718-482-1860

THE COCKPIT

CAPT. R. GENTILE & LT. J. GODFREY, U.S.A.A.F. DEBDEN, ENGLAND. 1944.

EARLY FALL, 1983. TWO DOLLARS.

THE COCKPIT

For M.C. & VISA Orders in New York Call 1-718-482-1860
Out of State M.C. & VISA Orders ONLY: Call TOLL FREE 1-800-354-5514
In N.Y.C. Visit Our Retail Store at 595 Broadway 1-212-925-5455

THE COCKPIT

VOLUME IX, ISSUE I.

The Cockpit of B-17G "Shoo Shoo Baby" The Air Force Museum, Dayton, Ohio. Photo: Dan Patterson (see p. 30 for the poster)

For M.C. & VISA Orders in New York Call 1-718-482-1860
Out of State M.C. & VISA Orders ONLY: Call TOLL FREE 1-800-354-5514
OUR NEW RETAIL STORE IS OPEN!!! VISIT US IN N.Y.C. AT 595 BROADWAY 1-212-420-1600

THE COCKPI

Sgt. Pilot Bill Arnot's North American B-25J "MITCHELL"

From GHOSTS: A Time Remembered Photo: Philip Makanna (see p.36)

For M.C. & VISA Orders in New York Call 1-718-482
Out of State M.C. & VISA Orders ONLY: Call TOLL FREE 1-
OUR NEW RETAIL STORE IS OPEN!!! VISIT US IN N.Y.C. AT 595 BRO

THE COCKPIT

VOLUME X, ISSUE III

A Quality Control Team Makes a Final Inspection of a 1200 H.P. 9 Cylinder Wright Cyclone R-1820 Engine before Installation in a Boeing B-17 "Fortress"

THE COCKPIT

"The Aviator" by J.C. Leyendecker: Oil on Canvas. Commissioned by Collier's Magazine, Sept. 1933. From the collection of Beverly and Ray Sacks.

THE COCKP